'A breathtaking overview of ph
and biology put together with
carefully explaining each new id
as and when it arises ... The te
Times Educational Supplement

'It is conscientiously and carefully written and strives to convey complex ideas clearly and simply ... well worth having' *Independent*

'From the interiors of atoms to dark matter, via evolution and the death of stars, John Gribbin explores science's power to explain the world. He obviously enjoys his subject and manages to find human angles on the trickiest concepts: for quantum theory he gets us to think about how a cashpoint only gives out denominations of £10s. This book will be hard to beat' *Focus*

'No one writes a better book on fundamental aspects of the Universe' *New Scientist*

'Any shortlist of the world's best popular science writers must include John Gribbin. His gift lies not just in eloquent and lucid presentation of complicated ideas, but in writing about them as if they were adventures' *Financial Times*

SOME BOOKS BY JOHN GRIBBIN

Galaxy Formation
White Holes
Timewarps
Genesis: The Origins of Man and the Universe
The Monkey Puzzle (*with Jeremy Cherfas*)
Spacewarps: Black Holes, White Holes, Quasars and the Universe
The Redundant Male (*with Jeremy Cherfas*)
In Search of Schrödinger's Cat: Quantum Physics and Reality
In Search of the Double Helix: Quantum Physics and Life
Weather (*with Mary Gribbin*)
In Search of the Big Bang: The Life and Death of the Universe
The Hole in the Sky
The Stuff of the Universe (*with Martin Rees*)
Winds of Change (*with Mick Kelly*)
Hothouse Earth: The Greenhouse Effect and Gaia
The Cartoon History of Time (*with Kate Charlesworth*)
The Matter Myth (*with Paul Davies*)
Stephen Hawking: A Life in Science (*with Michael White*)
Too Hot to Handle?: The Greenhouse Effect (*with Mary Gribbin*)
In Search of the Edge of Time
In the Beginning: The Birth of the Living Universe
Albert Einstein: A Life in Science (*with Michael White*)
Being Human (*with Mary Gribbin*)
Time and Space (*with Mary Gribbin*)
Schrödinger's Kittens and the Search for Reality
Darwin: A Life in Science (*with Michael White*)
Fire on Earth (*with Mary Gribbin*)
Watching the Weather (*with Mary Gribbin*)
Companion to the Cosmos
Richard Feynman: A Life in Science (*with Mary Gribbin*)
Q is for Quantum: Particle Physics from A to Z
Watching the Universe
In Search of Susy

Fiction

The Sixth Winter (*with Douglas Orgill*)
Brother Esau (*with Douglas Orgill*)
Father to the Man
Ragnarok (*with D.G. Compton*)
Innervisions

John Gribbin has a Ph.D in astrophysics from the University of Cambridge, and is now Visiting Fellow in Astronomy at the University of Sussex. His many bestselling books include *In Search of Schrödinger's Cat, The Omega Point, In Search of the Big Bang, In the Beginning, Schrödinger's Kittens, Companion to the Cosmos* and *Q is for Quantum*. His books have been translated into many languages, and have won awards in both Britain and the United States. He also writes science fiction. John Gribbin lives in Sussex with his wife Mary, also a science writer.

Almost Everyone's Guide to Science

THE UNIVERSE, LIFE AND EVERYTHING

JOHN GRIBBIN
with Mary Gribbin

PHŒNIX

A Phoenix Paperback
First published in Great Britain by Weidenfeld & Nicolson in 1998
This paperback edition published in 1999 by Phoenix,
a division of Orion Books Ltd,
Orion House, 5 Upper St Martin's Lane,
London WC2H 9EA

A CIP catalogue record for this book is available from the British Library.

ISBN: 0 75380 769 6

Typeset by Selwood Systems, Midsomer Norton

Printed and bound in Great Britain by
The Guernsey Press Co. Ltd, Guernsey, C.I.

CONTENTS

Concern for man himself and his fate must always be the chief interest of all technical endeavours ... in order that the creations of our mind shall be a blessing and not a curse to mankind. Never forget this in the midst of your diagrams and equations.

Albert Einstein
Caltech, 1931

SCIENTIFIC NOTATION

When dealing with very large or small numbers, as we shall be in this book, it is convenient to use scientific shorthand to avoid writing out strings of zeroes. In this standard scientific notation, 10^2 means 100 (a 1 followed by two zeroes), 10^3 means 1000, and so on. 10^{-1} means 0.1, 10^{-2} means 0.01, and so on. This notation comes into its own when we are dealing with numbers like Avogadro's Number (see chapter one), which is written as 6×10^{23} in scientific notation, shorthand for 600,000,000,000,000,000,000,000.

One point to watch is the significance of making a seemingly small change in the power of ten involved. 10^{24}, for example, is ten times bigger than 10^{23}, and 10^6 is not half of 10^{12}, but one millionth (10^{-6} times) as big.

We also follow the scientific convention in which a billion is a thousand million, or 10^9.

INTRODUCTION

IF IT DISAGREES WITH EXPERIMENT IT IS WRONG

The fate of specialists in any one area of science is to focus more and more narrowly on their special topic, learning more and more about less and less, until eventually they end up knowing everything about nothing.

It was in order to avoid such a fate that, many years ago, I chose to become a writer about science, rather than a scientific researcher. The opportunity this gave me to question real scientists about their work, and to report my findings in a series of books and articles, enabled me to learn less and less about more and more, although as yet I have not quite reached the stage of knowing nothing about everything. After thirty years of this, and many books focusing on specific aspects of science, it seemed a good idea to write a general book, giving a broad overview of science, while I am still at the stage of knowing a little bit about most things scientific.

Usually, when I write a book the target audience is myself – I write the book about, say, quantum physics, or evolution, that I wish somebody else had written for me so that I would not have had to go to the trouble of finding things out for myself. This time I am writing for everybody else, in the hope that there will be something here for almost everyone to enjoy. If you know a little quantum physics (or even a lot), you may find here something you didn't know about evolution; if you know about evolution, you may find something new to you about the Big Bang, and so on.

So, although I am aware of the ghost of Isaac Asimov looking over my shoulder (I hope with approval) at such a wide-ranging project, this is not 'John Gribbin's Guide to Science', but a guide for almost everyone else. A guide not so much for fans of science and the cognoscenti but more a guide for the perplexed – anyone who is vaguely aware that science is important, and might even be interesting, but is usually scared off by the technical detail. You won't find such technicalities here (they have all been removed by my co-author, who has kept my wilder scientific extravagances in check and has ensured that what remains is intelligible to a layperson). What you will find is one person's view of how science stands at the end of the twentieth century, and how the different pieces fit together to produce a coherent, broad picture of the Universe and everything it contains.

The fact that the pieces do fit together in this way is something you might miss from focusing too closely on one aspect of science, such as the Big Bang or evolution, but it is an extremely important feature of science. Both evolution and the Big Bang (and all the rest) are based on the same principles, and you can't pick and choose which bits of the scientific story you are going to accept.

I often receive communications from people who, for one reason or another, cannot accept the special theory of relativity, which tells us that moving clocks run slow and moving rulers shrink. Sometimes these people struggle desperately to find a way around this, while still accepting everything else in science. But you cannot. The special theory of relativity does not stand in isolation, as a theory about moving clocks and rulers, but comes into our understanding of, for example, the way mass is converted into energy to keep the Sun shining, and how electrons behave inside atoms. If you threw away the bits of the theory that seem to refute common sense, you would be left with no explanation of why the Sun is shining or of the periodic table of the elements. And this is just one example.

As I hope this book will make clear, *everything fits together* in the modern scientific world view. This scientific world view is the greatest achievement of the human intellect, and the power of that achievement stands out more clearly from a look at the broad picture than it does from too close attention to any one detail.

There are two remarkable, interconnected features of the scientific world view that are often overlooked, but are well worth pointing out. The whole thing has taken only about four hundred years to develop (starting from the time of Galileo, which seems as good a moment as any from which to date the beginning of modern scientific enquiry). And it can all be understood by a single human mind. Maybe we cannot all understand every bit of the scientific world view; but quite a few individual human beings can, even though people have such limited lifespans. And although it may take a genius to come up with an idea like the theory of evolution by natural selection, once that idea is formulated it can be explained to people of average intelligence – often provoking the initial response, 'How obvious; how stupid of me not to have worked that out for myself.' (This was, for example, pretty much Thomas Henry Huxley's reaction when he first read Charles Darwin's *Origin of Species*.) As Albert Einstein said in 1936, 'The eternal mystery of the world is its comprehensibility.'

The reason why the Universe is comprehensible to mortal minds is that it is governed by a small set of very simple rules. Ernest Rutherford, the physicist who gave us the nuclear model of the atom early in the twentieth century, once said that 'science is divided into two categories, physics and stamp-collecting'. He was not being entirely facetious, although he did have a genuine disdain for the other sciences, which made the fact that he was awarded the Nobel Prize for Chemistry (in 1908, for his work on radioactivity) deliciously inappropriate. Physics is the most fundamental of the sciences, both because it deals most directly with the simple rules that govern the Universe and the simple particles that everything in the Universe is made of, and because the methods of physics provide the archetype used by other sciences in developing their own parts of the world picture.

The most important of these methods is the use of what physicists call models. But even some physicists do not always appreciate just what their use of models is really all about so it is worth spelling this out before we think about applying the technique.

To a physicist, a model is a combination of a mental image of what some fundamental (or not so fundamental) entity is like, and a set of mathematical equations that describe its behaviour. For example, one model of the air that fills the room in which I am writing these words would regard every molecule of gas in the air as like a tiny, hard ball. There are accompanying equations which describe, at one level, how those little balls collide with one another and bounce off each other and the walls of the room and, at another level, how the average behaviour of very many of those little hard balls produces the pressure of the air in my room.

Don't worry about the equations – I shall largely ignore them in this book. But remember that good models always include the equations, and it is the equations that people work with in order to make predictions about the way objects behave – to calculate, perhaps, the way in which the pressure of the air in my room would change, other things being equal, if the temperature went up by ten degrees Celsius. The way to tell a good model from a bad one is to test it by experiment – in this case, warm the room up by ten degrees and see if the new pressure you measure matches the calculated pressure predicted by the model. If it does not, the model at best needs improvement and at worst may have to be discarded altogether.

Richard Feynman, the greatest physicist of the twentieth century, summed up the scientific process like this in a lecture he gave in 1964, using the word 'law' but forcefully making a point which applies equally well to models:

> In general we look for a new law by the following process. First we guess it. Then we compute the consequences of the guess to see what would be implied if this law that we guessed is right. Then we compare the result of the computation to nature, with experiment or experience, compare it directly with observation, to see if it works. If it disagrees with experiment it is wrong. In that simple statement is the key to science. It does not make any difference how beautiful your guess is. It does not make any difference how smart you are, who made the guess, or what his name is – if it disagrees with experiment it is wrong.

That is what science, and scientific models, are all about. *If it disagrees with experiment it is wrong.* But there is a more subtle point. Even if it *does* agree with experiment, that does not mean that a model is 'right' in the sense of being some eternal, universal Deep Truth about the nature of the thing that is being studied. Just because molecules can be treated as little, hard balls for the purpose of calculating the pressure of gas in a room, this does not mean that the molecules *are* little, hard balls – it means that under certain circumstances they behave *as if* they were little, hard balls. Models work within clear-cut – usually – limits, and outside those limits they may have to be replaced by other models.

To make this clear, let's take a different perspective on the image of the molecules of gas in the air in my room. Some of those molecules will be water vapour, and molecules of water, as every schoolchild knows, are made up of three atoms – two hydrogen and one oxygen, written as H_2O. For some purposes, a convenient model of the water molecule is two smallish hard balls (the hydrogen atoms) joined to a single larger hard ball (the oxygen atom) to make a V shape, with the oxygen at the vertex of the V.

For these purposes, the links between the atoms can be regarded as like little stiff springs, so that the atoms in the molecule can jiggle about, vibrating to and fro. This kind of vibration is associated with a characteristic wavelength of radiation – because the atoms carry electric charge (more of this later), if they are forced to vibrate in this way they will radiate microwave radio emission, and, alternatively, if the right kind of microwave radio emission is directed at the molecules they will vibrate in sympathy.

This is exactly what happens in a microwave oven. Microwaves tuned to the wavelengths that make water molecules vibrate fill the oven and make water molecules in the food vibrate, so that they absorb energy and heat up the food. This behaviour is not only seen in the kitchen or the laboratory – it is by studying the microwave radio emission coming from clouds of gas in space that astronomers detect the presence of water molecules in space, along with many other molecules.

So if you are a radio astronomer looking for molecules in space, or an electrical engineer designing a microwave oven, the stick and ball model of a molecule of water is a good one, provided the sticks joining the atoms are a little bit springy. You no longer regard the whole molecule as a single hard sphere; but you do regard the individual atoms, such as the oxygen atoms, as individual hard spheres.

A chemist analysing the composition of a substance would have yet another perspective. If you want to know which kinds of atoms are present in a substance, one way to find out is to study the light that they radiate when they get hot. Different kinds of atoms radiate different colours, very sharply defined lines in the rainbow spectrum of light – one of the most familiar examples is the bright orange-yellow colour of streetlights which contain compounds of sodium. It is the atoms of sodium (in this case, excited by an electric current, rather than by heat) which radiate this particular colour of light.

The model used to describe how this light is produced sees an atom not as a single hard sphere but as a tiny central nucleus (which can itself be thought of as a single hard sphere for now) surrounded by a cloud of tiny, electrically charged particles, called electrons. The central nucleus has a positive electric charge and the electrons each have negative electric charge, so that overall the atom has zero electric charge. The bright lines in the spectrum associated with a particular kind of atom are then explained in terms of the way electrons move in the outer part of the atom. What distinguishes one kind of atom from another, chemically speaking, is the number of electrons (8 for oxygen, just 1 for hydrogen, 11 for sodium); and because each kind of atom has its own unique arrangement of electrons, each kind of atom produces its own unique pattern of coloured lines in the spectrum.

I could go on, but the point is clear already. The model which treats molecules of air as little hard balls is a good one, because it works when you use it to calculate how pressure changes when temperature changes. The model which treats molecules as being made up of smaller hard spheres (atoms) held together like bunches of grapes is also a good one, because it works when you use it to calculate the way vibrating molecules make radio waves. And the model which treats atoms not as indivisible hard spheres but as tiny nuclei, surrounded by clouds of electrons, is also a good one, because it works when you use it to calculate the colour of light associated with a particular kind of atom.

None of the models is the ultimate Deep Truth, but they all have their part to play. They are tools which we use to help our imaginations to get a picture of what is going on, and to calculate things which we can test directly by measurement, such as the pressure of air in a room or the colour of light radiated by a hot substance.

Just as a carpenter would not use a chisel to do the same job as a mallet, so a scientist must choose the right model for the job in hand. When Feynman says 'if it disagrees with experiment it is wrong' he means if it disagrees with an *appropriate* experiment. The model of a molecule of water vapour as a single hard sphere does not allow for the possibility of the kind of vibration associated with microwaves, so it 'predicts' that water vapour will not radiate microwaves. This means that it is the wrong model to use if we are interested in microwaves. But it does *not* mean that it is the wrong model to use if we are interested in how the pressure of air in the room is affected by increasing the temperature.

Everything in science is about models and predictions, about finding ways to get a picture in your head of how the Universe works, and ways to make calculations that forecast what will happen in certain circumstances. The further we get from the ordinary world of everyday life, whether towards the very small scale or towards the very large scale, the more we have to rely on analogies: an atom is, under certain circumstances, 'like' a billiard ball; a black hole is, in some sense, 'like' a dent in a trampoline.

It would be tedious to keep qualifying the use of the various models in this way, and now that I have got this off my chest I shall largely refrain from doing so and trust you to remember the qualification that even the best model is only a good one in its own context, and that chisels should never be used to do the job of mallets. Whenever we describe something as being 'real', what we mean is that it is the best model to use in the relevant circumstances.

With that proviso, starting out from the scale of atoms, I shall take you down into the world of the very, very small, and then out into the Universe at large, giving the best modern understanding (the best model) of the nature of things on each scale. They are all true, in that they agree with experiment; they all fit together, like pieces in a jigsaw puzzle, to give a coherent picture of how the Universe, and everything in it, works; and it can all be understood, at least in outline, by an average human mind.

There is another feature of science, a view which I hold strongly, and which has shaped the structure of this book (and my entire career), but which is not necessarily shared by all scientists. To me, science is primarily an investigation of our place in the Universe – the place that people occupy in a world which ranges from the tiniest subatomic particles to the furthest reaches of space and time. We do not exist in isolation, and science is a human cultural activity, not a purely dispassionate striving after truth, no matter how hard we might try. It is all about where we came from, and where we are going. And it is the most exciting story ever told.

John Gribbin
December 1997

CHAPTER ONE

ATOMS AND ELEMENTS

In 1962, in a series of lectures given for undergraduates at Caltech, Richard Feynman placed the atomic model at the centre of the scientific understanding of the world. As he expressed it:

> If, in some cataclysm, all of scientific knowledge were to be destroyed, and only one sentence passed on to the next generations of creatures, what statement would contain the most information in the fewest words? I believe it is the *atomic hypothesis* (or the atomic *fact*, or whatever you wish to call it) that *all things are made of atoms – little particles that move around in perpetual motion, attracting each other when they are a little distance apart, but repelling upon being squeezed into one another.* In that one sentence, you will see, there is an *enormous* amount of information about the world, if just a little imagination and thinking are applied.

The emphasis is Feynman's, and you can find the whole lecture in his book *Six Easy Pieces.*[1] In the spirit of Feynman, we start our guide to science with atoms. It is often pointed out that the idea of atoms, as the ultimate, indivisible pieces of which matter is composed, goes back to the time of the Ancient Greeks, when, in the fifth century BC, Leucippus of Miletus and his pupil Democritus of Abdera argued the case for such fundamental entities. In fact, although Democritus did give atoms their name (which means 'indivisible'), this is something of a red herring. The idea wasn't taken seriously by their contemporaries, nor by anyone else for more than two thousand years. The real development of

[1] Details of books mentioned in the text are given in the Bibliography.

the atomic model dates from the end of the eighteenth century, when chemists began the modern investigation of the properties of the elements.

The concept of elements – fundamental substances from which all the complexity of the everyday world is made up – also goes back to the early Greek philosophers, who came up with the idea that everything is made up of different mixtures of four elements – air, earth, fire and water. Apart from the name 'element', and the idea that an element cannot be broken down into any simpler chemical form, there is nothing left of the Greek idea of elements in modern chemistry, which builds from the work of Robert Boyle in the middle of the seventeenth century.

Boyle was the first person to spell out the definition of an element as a substance that could combine with other elements to form compounds, but which could not be broken down into any simpler substance itself. Water, for example, is a compound which can be broken down chemically into its component parts, oxygen and hydrogen. But oxygen and hydrogen are elements, because they cannot be broken down further by chemical means. They are not made of other elements. The number of known elements increased as chemists devised new techniques for breaking compounds up; but by the nineteenth century it was becoming clear which substances really were indivisible.

The breakthrough in understanding the way elements combine to make compounds came when John Dalton revived the idea of atoms at the beginning of the nineteenth century. He based his model on the discovery that for any particular compound, no matter how the compound has been prepared, the ratio of the weights of the different elements present is always the same. For example, in water the ratio of oxygen to hydrogen is always 8:1 by weight; in calcium carbonate (common chalk) the ratio of calcium to carbon to oxygen is always 10:3:12 by weight.

Dalton's explanation was that each kind of element is composed of one kind of identical atoms, and it is the nature of these atoms that determines the properties of the element. On this picture, the key distinguishing feature that makes it possible to tell one kind of atom from another is its weight. When two or more elements combine, it is actually the atoms of the different elements that join together, to make what are now known as molecules. Each molecule of a compound contains the same number of atoms as every other molecule of the compound, each with the same numbers of atoms of each of the elements involved in each molecule. A molecule of water is made up of two hydrogen atoms and one oxygen atom (H_2O); a molecule of calcium carbonate is made up of one atom of calcium, one atom of carbon, and three atoms of oxygen ($CaCO_3$). And we now know that in some elements the atoms can join together to make molecules, without other elements being involved. The oxygen in the air that we breathe, for example, is made up of di-atomic molecules, O_2 – these are not regarded as compounds.

Dalton's atomic model was a huge success in chemistry, but throughout the nineteenth century some scientists regarded it only as a useful trick, a way to calculate the way elements behaved in chemical reactions, but not a proof that atoms are 'real'. At the same time, other scientists were finding increasingly compelling evidence that atoms could be regarded as real entities, little hard balls that attract each other when some distance apart but repel when pushed together.

One line of attack stemmed from the work of Amadeo Avogadro (who was, incidentally, the person who showed that the combination of atoms in a molecule of water is H_2O, not HO). In 1811 Avogadro published a paper in which he suggested that equal volumes of gas at the same temperature and pressure contain equal numbers of atoms. This was before the idea of molecules was developed, and we would now say that equal volumes of gas at the same temperature and pressure contain equal numbers of molecules. Either way, though, what matters is that Avogadro's model envisaged equal numbers of little hard spheres bouncing around and colliding with one another in a box of gas of a certain size under those conditions, whether the gas was oxygen, or carbon dioxide, or anything else.

The idea behind this is that in a box of gas there is mostly empty space, with the little hard balls whizzing about inside the box, colliding with one another and with the walls of the box. It doesn't matter what the little balls are made of – as far as the pressure on the walls of the box is concerned, all that matters is the speed of the particles and how often they hit it. The speed depends on the temperature (higher temperature corresponds to faster movement), and the number of hits per second depends on how many little hard balls there are in the box. So at the same temperature, pressure and volume, the number of particles must be the same.

This kind of model also explains the difference between gases, liquids and solids. In a gas, as we have said, there is mostly empty space, with the molecules hurtling through that space and colliding with one another. In a liquid there is no empty space, and the molecules can be envisaged as touching one another, but in constant movement, sliding past one another in an amorphous mass. In a solid the movement has all but stopped, and the molecules are locked in place, except for a relatively gentle jiggling, a kind of molecular running on the spot.

Avogadro's idea wasn't taken very seriously at the time (not even by Dalton). But at the end of the 1850s it was revived by Stanislao Cannizzaro, who realised that it provided a way of getting a measure of atomic and molecular weights. If you can find the number of molecules in a certain volume of one particular gas at a set temperature and pressure (the standard conditions are usually chosen as zero degrees Celsius and one standard atmospheric pressure), then you know the number of molecules present for any gas under those conditions. In order to find out how much each molecule weighs, you just have to weigh the gas and divide by that number.

For these standard conditions, you can choose a volume of gas which corresponds to two grams of hydrogen (two grams, not one, because each molecule of hydrogen contains two atoms, H_2). It works out as just over thirteen litres of gas. The number of molecules in such a volume is called Avogadro's Number. The same volume of oxygen under the same conditions weighs thirty-two grams, and chemical evidence tells us that there are two oxygen atoms in each molecule. But it contains the same number of molecules as two grams of hydrogen. So we know, immediately, that one oxygen atom weighs sixteen times as much as one atom of hydrogen. This was a very useful way to determine relative atomic and molecular weights; but working out actual weights depends on knowing Avogadro's Number itself, and that was harder to pin down.

There are several different ways to tackle the problem, but you can get some idea of the way it might be done from just one example, a variation on the theme used by Johann Loschmidt in the mid-1860s. Remember that in a gas there is a lot of empty space between the molecules, but in a liquid the molecules are touching one another. Loschmidt could calculate the pressure of gas in a container (under standard conditions) from Avogadro's Number, which determines the average distance travelled by the molecules between collisions (the so-called 'mean free path'), and the fraction of the volume of the gas actually occupied by the molecules themselves. And he could find out how much empty space there was in the gas by liquefying it and measuring how much liquid was produced – or, indeed, by using measurements of the density of liquid oxygen and liquid nitrogen that other people had carried out. Since the particles are touching in a liquid, by subtracting the volume of the liquid from the volume of the gas he could find out how much empty space there was in the gas. So by adjusting the value of Avogadro's Number in his pressure calculations to match the measured pressure, he could work out how many molecules were present.

Because the densities of liquid nitrogen and liquid oxygen used in his calculations were not as accurate as modern measurements, Loschmidt's figure for Avogadro's Number, derived in 1866, came out a little on the low side at 0.5×10^{23}. Using a different technique, Albert Einstein came up with a value of 6.6×10^{23} in 1911. The best modern value for the number is 6.022045×10^{23} or, in everyday language, just over six hundred thousand billion billion. This is the number of atoms in one gram of hydrogen, sixteen grams of oxygen, or in the gram equivalent of the atomic weight of any element. So each atom of hydrogen weighs 0.17×10^{-23} grams, and so on. Each molecule of air is a few hundred millionths of a centimetre across. At 0°C and one atmosphere pressure, one cubic centimetre of air contains 4.5×10^{19} molecules; the mean free path of a molecule of air is thirteen millionths of a metre, and an oxygen molecule in the air at that temperature travels at just over 461 metres per second (roughly

17,000 km per hour). So each molecule is involved in more than 3.5 billion collisions every second, producing the averaged-out feeling of a uniform pressure on your skin or on the walls of the room.

In fact, the kinetic theory of gases was first proposed by Daniel Bernoulli, as long ago as 1738. He was inspired by the work of Robert Boyle, in the middle of the seventeenth century. Boyle had discovered that when a gas is compressed (for example, by a piston) the volume of the gas changes in inverse proportion to the pressure – double the pressure and you halve the volume. Bernoulli explained this in terms of the kinetic theory, and also realised that the relationship between the temperature of a gas and its pressure (when you heat a gas its pressure increases, other things being equal) could also be explained in terms of the kinetic energy (energy of motion) of the little particles in the gas – heating the gas makes the particles move faster, so they have a greater impact on the walls of the container. But he was way ahead of his time. In those days most people who thought about heat at all thought that it was related to the presence of a kind of fluid, called caloric, which moved from one substance to another. Bernoulli's version of the kinetic theory made no impact at all on science at the time.

The kinetic theory was rediscovered twice (first by John Herapath in 1820, then by John Waterston in 1845), and ignored each time, before it finally became accepted by most scientists in the 1850s, largely as a result of the work of James Joule. A complete mathematical version of kinetic theory (a complete model) emerged in the 1860s, largely thanks to the work of Rudolf Clausius, James Clerk Maxwell, and Ludwig Boltzmann. Because this model deals with the averaged-out statistical behaviour of very large numbers of particles, which interact with one another like tiny billiard balls, bouncing around in accordance with Newton's laws of mechanics, it became known as statistical mechanics.

This is an impressive example of the way in which physical laws can be applied in circumstances quite different from those which were being investigated when the laws were discovered, and highlights the important difference between a law and a model. A law, like Newton's law of gravity, really is a universal truth. Newton discovered that every object in the Universe attracts every other object in the Universe with a force that is proportional to one over the square of the distance between the two objects. This is the famous 'inverse square law'. It applies, as Newton pointed out, to an apple falling from a tree and to the Moon in its orbit, in each case being tugged by the gravity of the Earth. It applies to the force holding the Earth in orbit around the Sun and to the force which is gradually slowing the present expansion of the Universe. But although the law is an absolute truth, Newton himself had no idea what caused it – he had no model of gravity.

Indeed, Newton specifically said in this context *hypotheses non fingo* ('I do not make hypotheses'), and did not try to explain why gravity obeyed an inverse square law. By contrast, Einstein's general theory of relativity provides a model which automatically produces an inverse square law of gravity. Rather than overturning Newton's ideas about gravity, as some popular accounts suggest, the general theory actually reinforces them, by providing a model to explain the law of gravity (it also goes beyond Newtonian ideas to describe the behaviour of gravity under extreme conditions; more of this later).

In order to be a good model, any model of gravity must, of course, 'predict' an inverse square law, but that doesn't mean that such a model is necessarily the last word, and physicists today confidently expect that one day they will develop a quantum theory of gravity that goes beyond Einstein's theory. If and when they do, though, we can be sure of one thing – that new model will still predict an inverse square law. After all, whatever new theories and models physicists come up with, the orbits of the planets around the Sun will still be the same, and apples won't suddenly start falling upwards out of trees.

Gravity, as it happens, is a very weak force, unless you have a lot of matter around. It takes the gravity of the entire Earth, pulling on an apple, to break the apple free from the tree and send it falling to the ground. But a child of two can pick the apple up from the ground, overcoming the pull of gravity. For atoms and molecules, rattling around in a box of gas, the gravitational forces between the particles are so tiny that they can be completely ignored. What matters here, as the nineteenth-century developers of statistical mechanics realised, are the other laws discovered by Newton – the laws of mechanics.

There are just three of these Newton's laws, which are so familiar today that they may seem like obvious common sense, but which form the foundations of all of physics. The first law says that any object stays still, or moves at a steady speed in a straight line (at constant velocity), unless it is pushed or pulled by a force. This isn't quite everyday common sense, because if you set something moving here on Earth (if I kick a ball, for example) it soon stops moving, because of friction. Newton's insight was to appreciate how things behave when there is no friction – things like rocks moving through space or, indeed, atoms whizzing about in a box of gas (incidentally, although he never developed a kinetic theory of gases, Newton was himself a supporter of the atomic model, and wrote of matter being made up of 'primitive Particles ... incomparably harder than any porous bodies compounded of them; even so very hard, as never to wear out or break in pieces').

Newton's second law says that when a force is applied to an object it accelerates, and keeps on accelerating as long as the force is applied (acceleration means a change in the speed of an object, or the direction in which it is moving or both; so the Moon is accelerating around the Earth, even though its speed stays much the same, because its direction is constantly changing). The acceleration produced by the force depends on the strength of the force divided by the mass of the object (turning this around, physicists often say that the force is equal to the mass times the acceleration). This does match up with common sense – it is harder to push objects around if they have more mass. And Newton's third law says that when one object exerts a force on another there is an equal and opposite reaction back on the first object. When I kick a ball (or, if I am foolish enough to do so, a rock), the force my foot exerts on the ball (or rock) makes it move, and the equal and opposite force the object exerts on my foot can be clearly felt.

More subtly, just as the Earth tugs on the Moon through

gravity, so there is an equal and opposite force tugging on the Earth. Rather than the Moon orbiting the Earth, we should really say that they each orbit around their mutual centre of gravity – but the Earth is so much more massive than the Moon that, as it happens, this point of balance actually lies below the surface of the Earth. Strictly speaking, too, the equality of action and reaction means that when the apple is falling to the ground the whole Earth, tugged by the apple, moves an infinitesimal amount 'up' to meet the apple. It is Newton's third law that explains the recoil of a gun when it is fired and the way a rocket works by throwing matter out in one direction and recoiling in the opposite direction.

These three laws apply to the Universe at large, which is where Newton applied them to explain the orbits of the planets, and to the everyday world, where they can be investigated by doing experiments such as rolling balls down inclined planes and measuring their speed, or by bouncing balls off one another. But, because they are, indeed, universal laws they also apply on the scale of atoms and molecules, providing the mechanics which, as I have mentioned, is the basis of statistical mechanics and the modern kinetic theory of gases – even though the modern kinetic theory was developed nearly two centuries after Newton came up with his three laws of mechanics, and he had never applied his laws in this way. Newton did not really invent the laws at all – they are laws of the Universe, and they operated in the same way before he wrote them down, just as they operate in places he did not happen to investigate.

There are good reasons why the kinetic theory, and statistical mechanics, were at last taken on board by scientists in the middle of the nineteenth century. By that time, the ground had been prepared by the study of thermodynamics (literally, heat and motion), which was of immense practical importance in the days when the industrial revolution in Europe was being driven by steam power.

The principles of thermodynamics can also be summed up in three laws, and they have a wide-ranging importance which applies across science (indeed, across the Universe), not just to the design and construction of efficient steam engines. The first law of thermodynamics is also known as the law of conservation of energy, and says that the total energy of a closed system stays the same. The Sun is not a closed system, and is pouring energy out into space; the Earth is not a closed system, and it receives energy from the Sun. But in a process like a chemical reaction taking place in an insulated test tube, or in the processes involved in statistical mechanics where little hard particles bounce around inside a box, the total amount of energy is fixed. If one fast-moving particle carrying a lot of kinetic energy collides with a slow-moving particle which has little kinetic energy, the first particle will probably lose energy and the second particle will gain energy. But the total energy carried by the two particles before and after the collision will be the same.

Since Einstein came up with the special theory of relativity early in the twentieth century, we know that mass is a form of energy, and that under the right circumstances (such as inside a nuclear power station, or at the heart of the Sun) energy and mass can be interchanged. So today the first law of thermodynamics is called the law of conservation of mass–energy, not just the law of conservation of energy.

The second law of thermodynamics is arguably the most important law in the whole of science. It is the law which says that things wear out. In terms of heat – the way the law was discovered in the days of steam engines – the second law says that heat will not flow from a colder place to a hotter place of its own volition. If you put an ice cube in a cup of hot tea, the ice melts and the tea gets cooler – you never see a cup of lukewarm tea in which the tea gets hotter while an ice cube forms in the middle of the cup, even though such a process would not violate the law of conservation of energy. Another manifestation of the second law is the way that a brick wall, left undisturbed, will be worn down and crumble away, while a pile of bricks, left undisturbed, will never assemble themselves into a brick wall.

In the 1920s the astrophysicist Arthur Eddington, slightly tongue-in-cheek, summed up the importance of the second law in his book *The Nature of the Physical World*:

> The second law of thermodynamics holds, I think, the supreme

> position among the laws of Nature. If someone points out to you that your pet theory of the universe is in disagreement with Maxwell's equations – then so much the worse for Maxwell's equations. If it is found to be contradicted by observation – well, these experimentalists do bungle things sometimes. But if your theory is found to be against the second law of thermodynamics I can give you no hope; there is nothing for it but to collapse in deepest humiliation.

The second law is also related to the concept known as entropy, which measures the amount of disorder in the Universe, or in a closed part of the Universe (such as a sealed test tube in the laboratory). Entropy in a closed system cannot decrease, so any change in the system moves it towards a state of higher entropy. The 'system' of an ice cube floating in a cup of tea has more order (less entropy) than a cup of lukewarm tea, which is why the system shifts from the ordered state to the disordered state.

The Universe as a whole is a closed system, so the entropy of the whole Universe must be increasing. But the Earth, as I have pointed out, is not a closed system and receives a continuous input of energy from the Sun. It is this supply of energy from outside that makes it possible for us to create order out of disorder locally (building houses out of piles of bricks, for example); the decrease in entropy associated with all the processes of life on Earth is more than compensated for by the increase in entropy inside the Sun as a result of the processes which make the energy we feed off.

In case you are wondering, the same sort of thing happens on a smaller scale when we cool down the inside of a refrigerator and make ice cubes. We have to use energy to pump heat out of the refrigerator, and this process increases the entropy of the Universe much more than the decrease in entropy that is produced inside the cold refrigerator as a result. If you left a fridge in a closed room, with the door of the fridge open and the motor running, the room would get *hotter*, not colder, because the energy being wasted by the motor getting hot would be more than the cooling effect of the open fridge.

The third law of thermodynamics has to do with the familiar everyday concept of temperature, something we have been taking for granted in the discussion so far. Although I have only discussed the relationship between heat and entropy in general terms, there is, in fact, a precise mathematical relationship between the two quantities, and this shows that as objects become cooler it is harder and harder to get energy out of them. This is pretty obvious in an everyday context – the very reason why steam engines were so important in the industrial revolution is because hot steam could be made to do useful work, driving pistons in and out or making wheels go round and round. You *could*, if you really wanted to, make a kind of toy engine that would push pistons in and out using a much cooler gas (perhaps carbon dioxide), but it would not be very effective.

In the 1840s, William Thomson (later Lord Kelvin) developed these thermodynamic ideas into an absolute scale of temperature, with zero on this scale being the temperature at which no more heat (or any energy) can ever be extracted from an object. This absolute zero is fixed by the laws of thermodynamics (so Thomson could calculate it mathematically, even though he could never cool anything that far), and is at –273°C. It is now known as 0 K, after Kelvin. On the Kelvin scale of temperature the units are each the same size as the degrees on the Celsius scale, so that ice melts at 273 K (there is no 'degree sign' in front of the K). The third law of thermodynamics says that you can never cool anything to 0 K, although if you try hard enough you can (in principle) get as close as you like to absolute zero. An object at zero Kelvin would be in the lowest possible energy state that it could achieve, and no energy could be extracted from it to do work.

Some wag has summed up the three laws of thermodynamics in everyday terms:

1. You can't win.
2. You can't even break even.
3. You can't get out of the game.

The success of the kinetic theory and statistical mechanics helped to convince many physicists that atoms were real. But right up until the end of the nineteenth century, many chemists still regarded the notion of atoms with suspicion. This looks very odd to us, because by the end of the 1860s (just about the time that the kinetic theory was proving so successful), a pattern had been discovered in the properties of the different elements – a pattern which is now explained entirely in terms of the properties of atoms.

An early attempt to rank the elements in order of their atomic weights (with the weight of hydrogen, the lightest element, set as 1 unit) was made by Jöns Berzelius in the 1820s (even though he never accepted Avogadro's hypothesis), but this never caught on. The real breakthrough came after 1860, when Cannizzaro revived Avogadro's idea and convinced many of his colleagues that it worked, and that the atomic weight was a useful concept in chemistry. Even then, it took time for the full importance of the breakthrough to be appreciated. The big discovery was that if the elements were arranged by atomic weight, elements with similar chemical properties were found at regular intervals in the list – the element with atomic weight 8 has similar properties to the one with weight 16 (and also to element 24), the element with atomic weight 17 has similar properties to the one with weight 25, and so on.

It didn't take much imagination to go from this discovery to the idea of writing out a list of the elements in a table, with the elements with similar properties arranged underneath each other in a set of vertical columns. In the early 1860s, the French chemist Alexandre Beguyer de Chancourtois and the British chemist John Newlands each independently came up with versions of this idea, but their work was ignored. Worse, Newlands' idea was made fun of by his contemporaries, who said that listing the elements in order of atomic weight made no more sense than listing them in alphabetical order. This was a breathtaking piece of arrant (and arrogant) nonsense, as the alphabet is an arbitrary human convention, while the weights of the atoms are a fundamental physical property, but the comment shows how far chemists were from accepting the reality of atoms in the mid-1860s.

Even when the idea finally began to catch on, there was an element of controversy. At the end of the 1860s, the German Lothar Meyer and the Russian Dmitri Mendeleyev each independently – and each unaware of the work of Beguyer de Chancourtois and Newlands – came up with the idea of representing the elements in a periodic table (a grid rather like a chessboard) in which they were arranged in order of their atomic weights, with elements with similar chemical properties underneath one another in the table. But the result is known to this day as Mendeleyev's Periodic Table, with Meyer relegated to a footnote of history along with the other two pioneers of the idea, because Mendeleyev was bold enough to rearrange the order of the elements in the table slightly, to make sure that elements with similar chemical properties fell in the same vertical column, even if this meant a slight shuffling of the order of the atomic weights.

These changes really were slight – for example, tellurium has an atomic weight of 127.61, just a little bit more than iodine, which has an atomic weight of 126.91. Reversing the order of these two elements in his table enabled Mendeleyev to put iodine under bromine, which it closely resembles chemically, and tellurium under selenium, which it closely resembles chemically, instead of having tellurium under bromine and iodine under selenium. We now know that Mendeleyev was right to make these changes, because the weight of an atom is determined by the combined number of (protons plus neutrons) in the atom, while its chemical properties are related to the number of protons alone – more of this in the next chapter – but neither the proton nor the neutron were known in the nineteenth century so there was no way that Mendeleyev could have explained the physical basis for this slight reordering of the elements in a table based on atomic weights, and he relied on the chemical evidence of similarities.

Mendeleyev's boldest step, and the one which eventually led to the widespread acceptance of his periodic table as something related to the fundamental properties of the elements, (and not an arbitrary convention like the alphabet) was his willingness to leave gaps in the table where there was no known element with properties that 'belonged' in a certain place. By 1871 Mendeleyev had produced a table containing the sixty-three elements known at the time, showing the striking periodicity in which families of elements with atomic weights that are multiples of eight times the atomic weight of hydrogen have similar chemical properties to one another. But to make the pattern work, even after minor adjustments like swapping the positions of tellurium and iodine, he had to leave three gaps in the table, and he boldly predicted that new elements would be found with properties (which he specified) corresponding to the places of those gaps in his table. The three elements, with exactly the predicted properties, were discovered over the next fifteen years: gallium, in 1875; scandium, in 1879; and germanium, in 1886.

In the classic tradition of science ('if it disagrees with experiment then it is wrong') Mendeleyev had made a prediction, and it had been proved correct. This persuaded people that the periodic table was important, and as more new elements were discovered and each one was found to fit into Mendeleyev's table, acceptance of his ideas turned to enthusiasm. There are now ninety-two elements known to occur naturally on Earth, and more than twenty heavier elements which have been created artificially in particle accelerators. All of them fit into Mendeleyev's table, allowing for some improvements to the layout of the table which have been made in the twentieth century to take account of our modern understanding of the structure of atoms.

But even the success of Mendeleyev's periodic table in the last third of the nineteenth century still did not persuade everyone of the reality of atoms. The final acceptance of the 'atomic hypothesis', as Feynman called it, came only in the first decade of the twentieth century – and it came, in no small measure, thanks to the work of the man whose image towers over twentieth-century science, Albert Einstein.

Einstein would surely have approved of Feynman's comments about the importance of the atomic hypothesis. He consciously devoted his first efforts as a research scientist to various attempts at proving the reality of atoms and molecules, notably in his PhD thesis (completed in 1905), and then in a series of scientific papers looking at the puzzle in different ways, and coming up with different ways to get a value for Avogadro's Number. It is a sign both of the importance of the concept of atoms and the continuing reluctance of scientists to take the idea fully on board that a physicist with the insight of Einstein still felt the need to do this at the beginning of the twentieth century. And there is no doubt what was in his mind when he did all this work. As he later wrote to Max Born, 'my main purpose ... was to find facts which would attest to the existence of atoms of definite size'.

The first way Einstein approached the problem (in his PhD thesis) was to calculate the rate at which sugar molecules in a solution in water pass through a barrier called a semi-permeable membrane, and to compare his calculations (which depended on the size of the molecules and their mean free path) with the results of experiments carried out by other people. This is conceptually very similar to the way Loschmidt got a handle on Avogadro's Number using mean free paths and molecular sizes for gases. A key point about Einstein, though, is that he never was an experimenter, and relied on experimental results provided by other people. So, in 1905, the best value he got for Avogadro's Number, using this technique, was 2.1×10^{23} – not because there was anything wrong with his calculations but because the experimental results weren't quite accurate. It was this technique, using data from more accurate experiments, that gave him a value of 6.6×10^{23} in 1911.

But in 1905 Einstein already had another way to 'attest to the existence of atoms'. This involved a phenomenon known as Brownian motion, named after the Scottish botanist Robert Brown. In 1827 Brown had noticed that pollen grains suspended in water can be seen (using a microscope) to move about erratically in a zig-zag dance. At first, this surprising discovery was interpreted by some people as a sign that the pollen grains were alive and active; but it soon became clear that the same dancing movement occurred with tiny particles of dust which could not possibly be alive.

In the 1860s, several physicists speculated that the motion might be caused by the little particles being buffeted about by the molecules of the liquid in which they were suspended (the same kind of motion is seen, for example, in particles of cigarette smoke suspended in the air). But this idea came to nothing at the time, largely because they had guessed, incorrectly, that each jerk of the grain (each zig or zag) must be caused by the impact of a single molecule, and that would mean that the molecules had to be huge – comparable in size to the suspended particles, which was obviously wrong, even in the 1860s.

Einstein tackled the problem from the other end. He was convinced of the reality of atoms and molecules, and wanted to find ways to convince others. He realised that a small particle suspended in a liquid would be buffeted by the molecules of the liquid, and calculated the kind of buffeting that would be produced. Einstein didn't know much about the history of the study of Brownian motion (throughout his career, Einstein didn't read up much on the history of any subject he was interested in, preferring to work things out for himself from first principles – this is an excellent way to do physics, provided you are as clever as Einstein). So in his first paper on the subject he calculated the way particles suspended in a liquid *ought* to move, and only said, cautiously, that 'it is possible that the motions discussed here are identical with the so-called Brownian molecular motion'. Colleagues who read that paper quickly reassured him that what he had described mathematically was *exactly* the observed Brownian motion – so, in a sense, Einstein predicted Brownian motion and the experiments confirmed his predictions. He was particularly taken by the idea, still true, that we can directly see the motions responsible for heat by looking at little particles suspended in liquids through a microscope. As he put it in 1905, 'according to the molecular-kinetic theory of heat, bodies of microscopically visible size suspended in a liquid will perform movements of such magnitude that they can be easily observed in a microscope.'

The insight on which Einstein based his calculations was that even an object as small as a pollen grain is being buffeted on all sides by a very large number of molecules at any instant. When it moves, jerkily, in one direction, it is not because it has received a single large blow pushing it that way, but because there has been a temporary imbalance in the buffeting – a few more molecules hitting it on one side than the other at that particular moment. Einstein then used some neat mathematics to work out the statistics of this kind of buffeting, and to predict the kind of zigzag path the pollen grain (or whatever) would follow as a result, with each little movement taking place in a totally random direction. It turns out that the distance of the particle from its starting point increases in proportion to the square root of the time that has elapsed. It travels twice as far in four seconds as it does in one second (2 is the square root of 4), it takes sixteen seconds to travel four times as far as it does in one second (4 is the square root of 16), and so on. And the direction the particle ends up in is oriented at random to its starting point, whether you look after four seconds, sixteen seconds, or any other number of seconds. This is now known as a 'random walk', and the same kind of statistics carries over into many other areas of science – including the behaviour of radioactive atoms when they decay.

The link between Brownian motion and Avogadro's Number was clear to Einstein, and he suggested how experiments might be carried out to study the exact movements of particles suspended in liquids and work out Avogadro's Number from these studies. But, as ever, he didn't carry out the experiments himself. This time the experiments were done by Jean Perrin, in France. Perrin studied the way a suspension of particles in a liquid forms layers, with most of the particles near the bottom and fewer higher up. The few particles that get higher up in the liquid (in spite of being tugged down by gravity) do so because they are kicked upwards by Brownian motion, and the height they reach depends on the number of kicks they receive, which depends on Avogadro's Number.

In 1908 Perrin used this technique to find a value for Avogadro's Number very close to the value then being found by several different techniques, and his experiments, (combined with Einstein's predictions) are generally seen as marking the moment (less than a century ago!) when the idea of atoms could no longer be doubted. As Einstein wrote to Perrin in 1909, 'I had believed it to be impossible to investigate Brownian motion so precisely.' Perrin himself wrote in the same year:

> I think it is impossible that a mind free from all preconception can reflect upon the extreme diversity of the phenomena which thus converge on the same result without experiencing a strong impression, and I think that it will henceforth be difficult to defend by rational arguments a hostile attitude to molecular hypotheses.

We trust you are indeed convinced that the atomic model is a good one. But before we move on to look inside the atom we would like to share with you just one more of the diverse phenomena which point in the same direction – the blueness of the sky.

This story goes back to the work of John Tyndall in the 1860s but culminates, once again, in a piece of work by Albert Einstein. Tyndall realised that the reason why the sky is blue is because blue light is scattered more easily around the sky than red light. Light from the Sun contains all the colours of the rainbow (or spectrum), with red at one end of the spectrum and blue, indigo and violet light at the other end, with everything mixed together to form white light. Red light has longer wavelengths than blue light, which means (among other things) that it does not scatter as easily from small particles as the blue end of the spectrum. Tyndall's idea was that the scattering which makes the sky blue, bouncing blue light around from particle to particle across the whole sky, is caused by dust particles and droplets of liquid suspended in the air.

He wasn't quite right. This kind of scattering does explain why sunsets and sunrises are red – red light penetrates dust or haze near the horizon better than blue light does – but the 'particles' needed to scatter blue light right around the sky, so that it seems to come at us from all directions, have to be very small indeed. Several physicists in the late nineteenth and early twentieth centuries suggested that the scattering might be caused by the molecules of air themselves. But it was Einstein who carried out the definitive calculations and proved, in a paper he wrote in 1910, that the blueness of the sky is indeed caused by the scattering of light from molecules of air. And, yet again, Avogadro's Number can be derived from the calculation. So you don't even need a microscope to see evidence that molecules and atoms exist – all you need is to look at the blue sky on a clear day.

To put some of the numbers concerning atoms in perspective, remember that the molecular weight of any substance in grams contains Avogadro's number of molecules. So just thirty-two grams of oxygen, for example, contains just over 6×10^{23} molecules of oxygen. Later in this book, we shall be discussing the nature of the Universe at large. Our Sun and Solar System are part of a disc-shaped galaxy of stars, the Milky Way Galaxy, which contains a few hundred billion (a few times 10^{11}) stars, each roughly similar to the Sun. In the whole Universe, there are several hundred billion galaxies visible in principle to our telescopes, and a research project I was involved with at Sussex University showed that our Galaxy is slightly smaller than the average disc-shaped galaxy. So how many stars are there altogether? Multiplying the numbers up, a few times several is probably about 10, and $10^{11} \times 10^{11}$ is 10^{22} so we have 10×10^{22}, or 10^{23}, stars in all. In round terms, the number comes out as a bit less than Avogadro's number of stars; which means that there are a few times *more* molecules of oxygen in just thirty-two grams of the gas (thirteen litres at standard temperature and pressure) than there are stars in the entire visible Universe.

To put this in a human perspective, the maximum capacity of the human lungs is about six litres; so if you take a deep breath, you have more molecules of air in your lungs than there are stars in the visible Universe.

In order to fit so many molecules into such a small amount of matter, each molecule (and each atom) has to be very small. There are many ways to calculate the sizes of atoms and molecules; the simplest is to take the volume of a liquid or solid that contains Avogadro's Number of these particles (thirty-two grams of liquid oxygen, for example) and divide that number into the volume to find out how big each one is – an idea that goes right back to the work of Cannizzaro but which can now be done with much greater precision. When you do this, you find that all atoms are roughly the same size, with the largest (caesium) 0.0000005mm across. That means that it would take ten million atoms, side by side, to stretch across the gap between two of the points on the serrated edge of a postage stamp.

At the beginning of the twentieth century, the idea of such tiny entities was just becoming fully accepted. But over the next few decades many of the greatest achievements in physics came not just from studying the behaviour of atoms, but from probing the structure within atoms, going down to scales one ten-thousandth the size of the atom to study the nucleus, and then to still smaller scales to study the fundamental particles of nature – at least, they are thought to be the fundamental particles of nature now, at the end of the twentieth century. We can begin to get a picture of what goes on inside the atom by looking at the way electrons are arranged in the outer parts of atoms, and at how they interact with light.

CHAPTER TWO

INSIDE THE ATOM

Physicists had already begun to probe inside the atom – although at first they did not appreciate that that was what they were doing – in the second half of the nineteenth century, before the concept of atoms was fully accepted. It came through the study of electricity.

Physicists studying the properties of electricity wanted to investigate it in as pure a state as possible. Electricity going through a wire is affected by the properties of the wire; even electric discharges sparking, like artificial lightning, across a gap between two charged plates in the air are affected by the properties of air. What the physicists needed was a way to study electric discharges jumping across the gap between two charged electrodes in a vacuum, and this only became possible in the middle of the 1880s, when Johann Geissler invented a vacuum pump powerful enough to evacuate the air from a glass container until the pressure was only a few ten-thousandths of the pressure of air at sea level.

Several people began to carry out experiments that involved passing electricity through such evacuated tubes. The basic principle was that two wires (the electrodes) would be sealed into the glass tube, one at each end, with a gap between them. One wire (called the anode) was attached to a source of positive electricity (for example, the positive terminal of a battery) and the other (the cathode) was attached to a source of negative electricity (for example, the negative terminal of the same battery). It soon became clear that when an electric current was flowing in the circuit, something was emerging from the cathode and travelling across empty space to the anode. The 'something' made its presence known by creating a glow in the tiny residual amount of air in the tube (a forerunner of the modern neon tube), or by striking the glass wall of the tube and making it glow.

The phenomenon was investigated in great detail by William Crookes, who concluded that these cathode rays, as they were known, were molecules of gas from the air remaining in the tube, which had picked up negative electric charge from the cathode and then been repelled from the cathode and attracted towards the anode (like charges repel one another; opposite charges attract one another). But he was wrong. A simple calculation soon showed that even at the low pressure inside the tubes, the mean free path of a molecule of air (the average distance it would travel without hitting another molecule) was only about half a centimetre; yet the cathode rays travelled in a straight line for almost a metre in some of the larger tubes built at the time.

But the cathode rays did seem to be a stream of particles. In 1895, Jean Perrin (who we met in chapter one) showed that the mysterious rays are deflected sideways by a magnetic field in exactly the way that a stream of negatively charged particles would be, and he found that when the cathode rays hit a metal plate the plate became negatively charged. He began to devise experiments to work out the properties of the particles in a beam of cathode rays, but was pre-empted by the independent work of J. J. Thomson in England, who did it first.

Thomson devised a brilliant experiment to get a handle on the nature of cathode rays – and 'devised' is exactly the right word, because J. J., as he was always known, was notoriously clumsy, and the experiments he dreamed up had to be carried out by more dextrous people with a lighter touch. By pushing a beam of cathode rays one way with a magnetic field, and the other way with an electric field, Thomson's team was able to find the strength of each kind of field needed to keep the cathode rays moving in a straight line. From this he was able to work out the ratio of the electric charge on a single particle in the beam divided by its mass – in shorthand, e/m. This had to be the same for every particle in the beam, because they all moved together in the electric and magnetic fields.

This may not sound like much of a breakthrough. What you really want is to find *e* and *m* separately. But the crucial point was that, in 1897, Thomson found that *e/m* for the particles in cathode rays was about one thousand times as big as *e/m* for a so-called ionised hydrogen atom, the lightest charged particle known up to that time. An ionised hydrogen atom is a hydrogen atom which has lost one unit of electric charge (we now know that this is because it has lost an electron). Making a bold leap (not entirely justified at the time) Thomson assumed that the amount of negative electric charge on each of his cathode ray particles was equal in size (but opposite in sign) to the amount of positive charge on a hydrogen ion. In that case, the mass of one of his cathode ray particles must be only one-thousandth of the mass of a hydrogen atom.

Thomson announced his results to the Royal Institution in London in 1897, commenting that 'the assumption of a state of matter more finely subdivided than the atom of an element is a somewhat startling one'. It was so startling, indeed, that some members of the audience thought that he was pulling their legs, and it was only two years later, in 1899, that further experiments provided convincing proof that these little negatively charged particles, which soon became known as electrons, were real. But in their eagerness to celebrate anniversaries, most physicists now regard 1897 as the date of the discovery of the electron, and duly celebrated the centenary of that discovery in 1997.

So, at the beginning of the twentieth century, while some scientists (including Einstein) were still putting the finishing touches to the various proofs that atoms exist, other physicists were already trying to work out what the internal structure of the atom must be, and how electrons could be chipped off, or knocked out of, what had previously been thought of as indivisible entities.

One line of attack was followed in experiments devised by Thomson which involved studying the behaviour of positively charged ions. It was Thomson (or rather, his team) who showed that these ions do, indeed, behave exactly as if they are atoms from which some negative electric charge has been removed, leaving them with an overall positive charge – exactly what you would expect if negatively charged electrons had been kicked out of them.

During the second decade of the twentieth century this work led to the discovery that atoms of the same element do not necessarily all have the same mass. Thomson's team measured the ratio e/m for ions with the same amount of electric charge, and used this to measure their masses. They found that, for example, neon comes in two varieties, one of which has atoms each twenty times the mass of a single hydrogen atom, the other which has atoms with twenty-two times the mass of a hydrogen atom. Such different varieties of the same element are now known as isotopes. It is the existence of different isotopes of the same element that accounts for the fact that the average atomic weight of an element may not be a whole multiple of the atomic weight of hydrogen, and which explains why some of the elements in Mendeleyev's periodic table seem to be out of sequence, chemically speaking, when they are arranged by average atomic weights. For any pure isotope, though, the atomic weight is always an exact multiple of the weight of a hydrogen atom. This was a significant discovery in determining the nature of the structure within atoms.

The other main line of attack came from theorists trying to build a model to account for Thomson's discovery that atoms are made up of electrons and something else, and that the electrons can be knocked out of atoms. The first of these models was devised by Lord Kelvin in 1902 (Kelvin had been born William Thomson, but he was no relation to J. J.). This model of the atom envisaged it as a tiny sphere about one-tenth of a billionth of a metre (0.1×10^{-9} m or 0.1 nanometres) across, with a positive charge spread out evenly through the sphere, and electrons embedded in the sphere like cherries in a cherry bun. By that time, though, the discovery of radioactivity had provided physicists with a means to probe the structure of the atom – and, if it disagrees with experiment, then it is wrong.

The first evidence for energetic processes going on inside atoms had come in 1895, when Wilhelm Röntgen discovered X-rays. Like many of his contemporary physicists, Röntgen was experimenting with cathode rays at the time, and he noticed that when the stream of cathode rays strikes an object (even the glass wall of the cathode ray tube), it can make the object emit another kind of radiation. Röntgen just happened to have a device called a fluorescent screen lying on a bench in his lab near the cathode ray experiment, and noticed tell-tale flashes of light coming from the screen when the cathode ray experiment was running. The cause of these flashes was the mysterious X-radiation, given its name because in mathematics X is traditionally the unknown quantity.

We now know that X-rays are just like light, a form of electromagnetic radiation, but with much shorter wavelengths than visible light. In terms of understanding the structure of the atom, though, what matters is that the discovery of X-rays in 1895 stimulated other people to search for other forms of radiation from atoms, and in 1896 Henri Becquerel discovered that atoms of uranium spontaneously produce another kind of radiation. Two years later Ernest Rutherford showed that there are actually two kinds of this atomic radiation, which he called alpha rays and beta rays. A third form of radiation, gamma rays, was discovered later. Beta rays were found to be very fast-moving electrons, and gamma rays turned out to be electromagnetic radiation like X-rays, but with even shorter wavelengths. It was alpha rays, though, that unlocked the next secret in the investigation of the structure of the atom.

At the beginning of the twentieth century nobody knew what alpha rays were, except that they were particles carrying two units of positive charge. Rutherford showed that they have the mass of a helium atom (strictly speaking, in modern terminology the mass of one particular kind of helium atom, the isotope known as helium-4). They are the same as helium atoms from which two electrons have been removed – what are now known as helium nuclei. What mattered to Rutherford, though, was that they were fast-moving particles which could be fired at atoms to probe their structure.

In 1909 two physicists working under the direction of Rutherford, Hans Geiger and Ernest Marsden, made the astonishing discovery that when a beam of alpha particles was fired at a thin sheet of gold foil most of the particles went straight through the foil, but a few bounced back almost in the direction they had come from. This, Rutherford said later, was 'quite the most incredible event that has ever happened to me ... It was almost as incredible as if you fired a 15-inch shell at a piece of tissue paper and it came back and hit you'.

There was no way that this could be explained by the cherry-bun model of the atom. On that picture, the atoms would be wedged in side by side in the gold foil, with a uniform spread of density right across the foil. Alpha particles ought to be slowed down, passing through the serried ranks of atoms – like a rifle bullet being slowed down if it is fired through a tank of water – but there ought not to be any hard centres that the alpha particles could bounce off. So Rutherford had to come up with a better model of the atom, one that did agree with the experiments. He announced it in 1911, and it is the model that we learn about in school today.

In Rutherford's model of the atom, almost all of the mass of the atom is concentrated in a tiny central nucleus, which is positively charged, and this nucleus is orbited by the negatively charged electrons, very roughly in the same sort of way that the planets orbit around the Sun. And just as most of the Solar System is empty space so most of the volume occupied by an atom is empty space. Most alpha particles fired at any 'solid' object brush right through the electron clouds around the nuclei almost unimpeded; but just occasionally an alpha particle hits a nucleus head-on, and is deflected.

From the frequency with which these deflections occurred, Rutherford could work out the size of the nucleus, compared with the size of the atom. An atom is about 10^{-8} cm across; but the nucleus is only 10^{-13} cm across. It has one hundred-thousandth of the diameter of the whole atom, equivalent to the size of a pinhead compared with the dome of St Paul's cathedral in London. Because volume goes as radius (or diameter) cubed, this means that the proportion of solid matter to empty space in an atom is not 10^{-5}, but 10^{-15}. Only one millionth of a billionth of an atom is solid nucleus. And since everything on Earth is made of atoms, that means that your own body, and the chair you sit on, are each made up of a million billion times more empty space than solid matter. The only reason why your own body, and the chair you sit on, seem to be solid and impenetrable is because those tiny specks of matter are held together by electric forces that operate between the charged particles, the nuclei and the electrons.

Putting all of this together, we have a model of the atom that works very well, within certain limits. By 1919 Rutherford had discovered that sometimes when a fast-moving alpha particle hits a nucleus of nitrogen the nucleus changes to one of oxygen, and a hydrogen nucleus is ejected. This suggested that atomic nuclei contain particles which are equivalent to nuclei of hydrogen, and other experiments confirmed this. The particles were given the name protons. As each proton has a positive charge exactly equal in size to the negative charge on an electron, and overall atoms are electrically neutral, every atom has the same number of protons in its nucleus as there are electrons in the cloud outside the nucleus, so that the charges balance. For the first time, it became possible to explain chemistry in terms of the structure within the atom (*see* chapter four).

One obvious puzzle about Rutherford's model, though, was how all the positive charge in the nucleus could stay there without blowing itself apart (repelling itself) – at least in the cherry-bun model the negative and positive charges were mixed up and could hold each other in place. The obvious answer was that there might be a kind of neutral particle in the nucleus, acting as a kind of ballast to hold the positive stuff together. These particles, called neutrons, were only identified in 1932.

As well as helping to hold nuclei together, the presence of neutrons in nuclei explained the existence of isotopes. Neutrons have very nearly the same mass as protons, so they make a big contribution to the weight of an atom. The chemical behaviour of an element depends almost entirely, as we shall see in chapter four, on the number of electrons associated with each atom of the element. That is the same as the number of protons in the nucleus. But the atomic weight of an element depends on the combined number of (protons + neutrons) in each nucleus, so by having different numbers of neutrons in their nuclei, but the same number of protons, atoms of different isotopes can have different weights, but the same chemical properties.

Beta radiation is produced when an individual nucleus emits an electron, gaining one unit of positive charge in the process. At a more detailed level, we can say that one of the neutrons in the nucleus has emitted an electron, transforming itself into a proton (this is the process known as beta decay). Because this increases the number of protons in the nucleus, it is changed into the nucleus of another element. Similarly, in alpha decay an unstable nucleus emits a package of two protons and two neutrons, bound together as an alpha particle. Once again, the nucleus left behind is thereby transformed into a nucleus of another element.

Even with neutrons present in the nucleus to help hold things together, there is still a lot of positive charge present, trying to blow the nucleus apart. So physicists realised that there must be a previously unknown kind of force, now known as the strong nuclear force, which holds everything together in the nucleus. The strong nuclear force is quite different from the two forces we experience in the everyday world, gravity and electromagnetism. They are both long-range forces – anyone who has played with a magnet knows how the force seems to reach out across space and latch on to a nearby piece of steel, while gravity reaches out from the Earth to keep the Moon in its orbit (and beyond!). But the strong nuclear force has only a very short range. It is a force of attraction which operates on both neutrons and protons, pulling them towards each other. And at very short distances (over the volume of a nucleus) it is almost a hundred times stronger than the electric force, so it overwhelms the repulsion between the positively charged protons. But, unlike the electric force, the strength of the strong nuclear force drops off dramatically.

Protons and neutrons (collectively often referred to as nucleons) only feel the force if they are practically touching each other. So, for example, if (for whatever reason) a positively charged alpha particle can get a little way out from the nucleus, the strong nuclear force effectively disappears, and it is repelled by the positive charge on the remaining nucleus. You can make an analogy with a stiff spring that you squeeze in your hands, compressing it and overcoming the force that tries to make it expand; as soon as you let go, the spring springs out with full force. We'll have more to say about the forces of nature in the next chapter. In passing, though, it is worth mentioning that beta decay is also regarded as being caused by a force, which is called the weak nuclear interaction.

The picture that began to emerge following the work of Rutherford was of a tiny, positively charged nucleus with electrons 'in orbit' around it. It took until the 1930s for the neutron to be identified as an individual particle and for its properties to be studied, but in modern language an atom of common hydrogen consists of a single proton orbited by a single electron. There is an isotope of hydrogen, called deuterium or heavy hydrogen, in which each nucleus contains both a proton and a neutron, but the outer part of the atom still has only a single electron. The next element is helium, which has two protons in its nucleus. Without a neutron to hold them together, the strong force would not be effective, and two protons on their own would be repelled by their mutual positive charge. So even the simplest atom of helium has two protons and one neutron in its nucleus, and two electrons outside the nucleus. It is known, for obvious reasons, as helium-3. The most common form of helium (helium-4) actually has two protons and two neutrons in its nucleus, with two electrons outside. The helium-4 nucleus is the same as an alpha particle, and this is a particularly stable configuration.

So the elements build up. The most common isotope of carbon, carbon-12, has six neutrons and six protons in its nucleus, with six electrons outside. Thanks to this pleasing symmetry, and because it is easier to work with carbon than with helium, today atomic weights are defined so that the weight of a carbon-12 atom is exactly 12; but for anyone who isn't worried about the precise details you can still think of atomic weights as being measured in units where hydrogen has a mass of 1. The difference is small, but real, because the common hydrogen atom has only a single proton and no neutron in its nucleus, and the masses of protons and neutrons are slightly different from one another.

For successively heavier elements there is a tendency for there to be more neutrons that protons in the nucleus, because they are needed to hold the increasing amount of positive charge together. Iron-56, for example, has 26 protons and 30 neutrons in its nucleus, while uranium-238 has 92 protons but no less than 146 neutrons in each nucleus. Indeed, by the time we get to uranium the nucleus is getting so big, with so much positive charge on board, that the strong nuclear force barely has the strength to hold the nucleus together, even with the extra glue provided by the neutrons. Each nucleon only feels the attractive strong force from its immediate neighbours – but each proton feels the repulsive electric force of all the other 91 protons in the nucleus, which is why the number of naturally occurring elements stops at 92. Physicists can make heavier nuclei in particle accelerators, but they are unstable and blow themselves apart in a short time.

All of this discussion, though, deals with the nuclei of the elements. There is one big puzzle about the Rutherford model of the atom, which was clear even before people knew about the existence of neutrons in the nucleus. How could all the electrons (or even the single electron associated with the hydrogen atom) stay in orbit?

The problem is that when an electric charge is accelerated it radiates energy – electromagnetic radiation, such as light. As we mentioned in chapter one, even circular motion at a steady speed is acceleration, because the direction of the particle involved is constantly changing. It is this kind of acceleration that presses you to one side when you are in a car that goes round a corner at high speed. Experiments in the lab confirm that if you whirl an electric charge around in a circle it radiates energy. But remember the law of conservation of energy. If an electron orbiting in an atom radiates it must lose energy. The only source of energy it has to draw on is the energy of motion that keeps it orbiting around the nucleus. So an electron in orbit around a nucleus ought to radiate energy away and fall downwards, spiralling in until it fell into the nucleus itself. According to the laws of physics as they were understood at the beginning of the twentieth century (usually referred to as 'classical' physics), every atom ought to collapse in an instant, in a puff of radiation.

The only way around this problem was to invoke a new kind

of physics – a new kind of model of the atom. As ever, though, this does not mean that the old models had to be abandoned. There must be something different going on on the scale of atoms which stops them collapsing in this way; but classical physics is still a perfectly good description of what happens to an electric charge that is whirled around in the lab.

The person who made the crucial step forward in finding a way to stabilise the Rutherford atom was Niels Bohr, a Danish physicist who worked for a time with Rutherford in England. He brought the idea of quanta into atomic physics.

Bohr's model of the atom was published in 1913, after he had returned to Denmark. It was an unashamed mixture of classical ideas (notably the idea of an orbit) and the new quantum ideas, with the great benefit that it worked. It not only gave an insight into why electrons didn't fall in to nuclei, but it also explained the pattern of lines produced by each element and seen in the spectrum of light – in particular, the pattern associated with hydrogen, the simplest element. All of this was explained later, in the 1920s, in even more satisfactory detail by a complete quantum mechanical description of atoms and radiation, but Bohr's model is so easy to understand that it is still the one most of us learn about in school.

The basic idea of quantum physics dates back to work by Max Planck, which he announced in Berlin in 1900; but it took a long time for the full implications of that work to sink in.

Before 1900 the accepted model of light treated it as a form of electromagnetic wave, moving through space. Discoveries in the nineteenth century had shown that a changing electric field produces a magnetic field and that a changing magnetic field produces an electric field. This idea of fields is so important that we shall return to it in the next chapter, but the idea is familiar from games with toy magnets – the field of the magnet is the region around the magnet where it exerts its magnetic influence.

James Clerk Maxwell discovered the equations that describe a pair of changing fields moving together through space – the changing electric field produces a changing magnetic field, and the changing magnetic field produces a changing electric field, marching together in step. Together they make an electromagnetic wave. Light, radio, and other forms of electromagnetic wave are all described by Maxwell's equations. The energy of the wave is provided by the energy put into the system in the first place – the electric current that flows through a wire to make the hot glow of a light bulb, for example.

This was all fine and dandy, and Maxwell's equations were one of the greatest triumphs of nineteenth-century physics. But there was a snag. Treating light simply as a form of wave did not explain the kind of radiation that hot objects actually produce. You make the electromagnetic radiation, ultimately, by jiggling electric charges about (accelerating electrons themselves, in fact). And if you apply the same sort of statistical rules that apply to the way waves behave in the classical world – for example, the sound waves that you get when you pluck a guitar string – it turns out that an accelerated electron ought to produce a huge number of electromagnetic waves with very short wavelengths and hardly any with long wavelengths.

Planck found a way round the problem by treating the radiation as if it could only be emitted in chunks of a definite size – quanta. The amount of energy in each quantum of radiation is inversely proportional to its wavelength so, on this picture, a quantum corresponding to short wavelengths takes a lot more energy to make than a quantum corresponding to long wavelengths. The relationship between the energy that a quantum carries and its wavelength depends on a number now called Planck's constant, which can be determined from experiments. For a particular quantum, the energy that it carries is equal to Planck's constant divided by the wavelength.

If you make a lot of electrons jiggle about together by heating the material they are part of, only a few of them will have enough energy to make the high-energy, short-wavelength quanta, so only a little high-energy radiation will be released. There will be a lot more electrons with enough energy to make medium-sized quanta with medium wavelengths, so a lot of medium-energy radiation will be released. But although there are lots of electrons with enough energy to make low-energy, long-wavelength quanta, each quantum contributes only a tiny amount, so only a little low-energy radiation is released.

Adding all of the contributions together in the proper statistical manner, Planck found that a hot object should radiate most of its energy in the middle of a band of wavelengths, with less energy emitted both at shorter and at longer wavelengths. And as the object gets hotter there are more electrons with enough energy to make shorter wavelength radiation, so the balance shifts and the peak in the energy emission moves to shorter wavelengths. This is exactly what we see in real life – a red-hot poker is cooler than a poker glowing orange, and red light has longer wavelengths than orange light.

Planck's idea worked perfectly to explain the nature of the electromagnetic radiation from hot objects, and how the colour of the object changes as it gets hotter.[1] But the idea of light as a form of electromagnetic wave was so well established that at first nobody took on board the idea of each quantum of light being a real entity in its own right – what we now call a photon. The idea was that there was something about atoms and electrons that prevented them radiating light except in amounts of a certain size, not that light only existed in amounts of a certain size.

My favourite analogy is with the cash dispenser at the bank. The cash machine will only dispense money in multiples of £10. I can get out £20, or £60, or any other multiple of ten (provided there is enough in my account and in the machine). But I can never get out, say, £27.43, even if I have that much in my account, and even though such an amount of money does exist, because the machine doesn't work that way. In the same way, early in the twentieth century physicists thought that electromagnetic radiation existed in waves with any kind of wavelength, with any kind of energy, but that atoms were only 'allowed' to release light in packets with a certain amount of energy for each particular wavelength.

Over the next twenty-five years it gradually became clear that in some circumstances the light quanta – the photons – had to

[1] For historical reasons, this is known as 'black body' radiation, because the model also explains how radiation is absorbed by a black object. But a hot, radiating object can emit any colour of 'black body' radiation if it has the right temperature.

be treated as real entities, and that the idea of electromagnetic waves simply would not work under all circumstances. Extending the cash-machine analogy, money is, in fact, quantised – the basic unit of money is the penny, and you can never have (in cash) an amount of, say, 241.378 pennies, only 241 pennies or 242 pennies. What matters is that the unit of money (the 'quantum' penny) is so small that it is possible to have very nearly any amount of money. In this example, few adults would be bothered, or even notice, whether they actually had £2.41 or £2.42 in their pocket.

On the photon picture, light is like that. There are so many photons, each contributing its own tiny amount of energy to a beam of light, that they add up to give the appearance of a smooth and steady flow of radiation. How many? About a thousand billion (10^{12}) photons of sunlight fall on a pinhead each second on a sunny day; when you look at a faint star your eye receives a few hundred photons from that star each second.

The discovery that light behaved as if it were a stream of particles caused enormous confusion for a time, and bewilderment among physicists, who slowly learned to think of entities such as light as being both particle and wave, sometimes showing one face to the world and sometimes the other. A quantum entity such as a photon is neither a particle nor a wave. It is something we cannot comprehend in terms of our everyday experience, except by explaining its behaviour under some circumstances using the wave analogy to describe its character, and under other circumstances using the particle analogy to describe its behaviour. Whatever you do, don't waste time trying to fathom out what a quantum entity like a photon 'really is' – nobody knows what it *is*, only what it is *like*. To quote Feynman again:

> Do not keep saying to yourself, if you can possibly avoid it, 'But how can it be like that?' because you will go 'down the drain' into a blind alley from which nobody has yet escaped. Nobody knows how it can be like that.[2]

[2] *The Character of Physical Law*, p. 129

But this is getting slightly ahead of the 1913 version of the Bohr model of the atom, which only needed the idea of quantisation of the emission and absorption of radiation (not necessarily the quantisation of the radiation itself) to solve the puzzle of why electrons in atoms did not fall into the nuclei.

Bohr developed Rutherford's planetary model of the atom by saying that electrons could only occupy certain 'stable orbits' around the nucleus, each corresponding to a certain fixed amount of energy, a multiple of the basic quantum. But there were no in-between orbits, because they would correspond to fractional amounts of energy. An electron might jump from one orbit to another, emitting a quantum of energy if it was moving closer to the nucleus or absorbing a quantum of energy if the jump took it further out from the nucleus. But it could not spiral in steadily towards the nucleus.

So why didn't all the negatively charged electrons simply jump straight into the nucleus, attracted by its positive electric charge? Bohr added another ingredient to his model, arguing that each stable orbit around the nucleus in some sense only has room for a limited number of electrons. If the orbit was full up, then no matter how many electrons there might be in orbits with more energy they could not give up that excess energy and jump down into the occupied orbit. Equally, it was simply forbidden for the electrons in the lowest energy orbit to make the final jump into the nucleus itself. But if the lower orbit had room for it, an electron in a higher energy orbit could jump down into it, radiating one quantum of energy (what we would now call a photon corresponding to a particular wavelength of light) as it did so. In order to jump up from a lower energy orbit to a higher one, an electron would have to absorb precisely one quantum with the right amount of energy – the same amount of energy as it would radiate if it fell back to the lower orbit.

The image is rather as if each electron sits on one step of a staircase, and can only jump up or down by a whole number of steps, because there are no in-between steps to rest on – but the picture is complicated because the steps are not all the same height.

It seemed natural to assume that every atom of a particular

element behaves in the same way, with the same spacing of steps on the energy-level staircase available to its electrons, since atoms of the same element (strictly speaking, isotopes of the same element) are identical to one another. So what we actually see when we look at the light being radiated by a hot object is the combined effect of all the little packets of light produced by electrons jumping about on a vast array of identical energy staircases.

What we actually do see in the spectrum of light from a hot object is a series of bright lines, at sharply defined wavelengths. These could now be explained as wavelengths corresponding to the quantum of energy radiated when an electron made a particular jump down the energy staircase – a particular transition from one energy level to a lower one. The pattern of these lines in the spectrum is different for each element, and acts as a unique fingerprint which shows the presence of a particular element in the hot object. Hot hydrogen gas, for example, produces its own distinctive spectral fingerprint, different from the fingerprint of every other element, even its closest atomic relative, helium. And if you shine light through a cold gas, the identical pattern of dark lines shows up in the spectrum, where energy has been absorbed as the light passes through the gas, used up in making electrons jump from one energy level to a higher energy level.[3]

The power of spectroscopy in identifying the presence of different elements had been discovered in the nineteenth century, and developed by researchers such as Joseph Fraunhofer, Gustav Kirchoff and Robert Bunsen. The famous 'Bunsen burner' (which was not actually invented by Bunsen, although he did make good use of it) was a key tool in this work. When a substance is heated in the clear flame of the burner (for example, when a wire is dipped into a powder or liquid and then held in the flame), it produces light with a characteristic colour, caused by the bright

[3] When an electron is in a high-energy level (called an 'excited' state), and has a choice of two or more lower-energy levels to fall into, it chooses its destination entirely at random. This is one feature of a so-called 'quantum leap'. The other feature is that the leap is very small by the standards of the everyday world. The spaces between the energy steps are only big enough to allow the release (or absorption) of one photon of light at a time. So a quantum leap is a very small change, made entirely at random.

lines in the spectrum at the characteristic wavelengths – that is, colours – of the element being studied.

One of the most familiar everyday examples of this kind of process is the distinctive orange-yellow colour of many street-lights, caused by the presence of sodium. In this case, it is the energy of electricity passing through the gas in the light that moves the electrons in sodium atoms up to a higher energy level. When they fall back down the energy staircase, they emit light with sharply defined wavelengths, forming two bright lines in the yellow part of the spectrum. The characteristic sodium yellow is seen if, for example, ordinary salt (sodium chloride) is heated in the flame of a Bunsen burner, or even if the salt is thrown on a fire.

Every element has its own pattern of spectral lines, and in every case the pattern stays the same (although the intensity of the lines will change) even if the temperature changes. By comparing their laboratory studies of spectra with the lines seen in the light from the Sun and stars, spectroscopists were able to account for most of these lines in terms of the presence in the Sun and stars of elements known on Earth (this kind of study began in 1859, when Kirchoff found that there is sodium in the atmosphere of the Sun).

In a famous reversal of this procedure, the British astronomer Norman Lockyer explained lines in the solar spectrum that did not correspond to any known element on Earth by attributing them to an unknown element, which he dubbed helium (named from the Greek word for the Sun, *helios*). Helium was later discovered on Earth and found to have exactly the spectrum required to fit these particular solar lines.

The spectrum of hydrogen is particularly simple – we now know that this is because the hydrogen atom consists of a single proton associated with a single electron. The lines in the spectrum that provide the unique fingerprint of hydrogen are named after Johann Balmer, a Swiss schoolteacher who worked out a mathematical formula describing the pattern in 1884, and published it in 1885. Balmer's formula was so simple that it clearly contained some deep truth about the structure of the hydrogen atom. But nobody knew what – until Bohr came on the scene.

Bohr was not a spectroscopist and, although he had done some spectroscopy as an undergraduate, when he began working on the puzzle of the structure of the hydrogen atom the Balmer series did not immediately occur to him as the key with which to unlock the mystery. It was only when a colleague pointed out to him just how simple the Balmer formula really is that he appreciated its importance. This was in 1913, and it led directly to Bohr's model of the hydrogen atom, with the single electron able to jump from one energy level to another. The spacing between the energy levels depends on the size of Planck's constant, and this in turn determines the spacing between the lines in the Balmer series. So by using the observed spacing between the lines in the series, Bohr was able to calculate the spacing of the energy levels, and Balmer's formula could be rewritten, in a very natural way, to include Planck's constant. It was this work by Bohr, drawing on nineteenth-century physics, which showed definitively that the hydrogen atom contains exactly one electron.

Bohr's model of the atom explained both the appearance of bright lines in the emission spectrum of an element, and the presence of dark lines at the same wavelengths in its absorption spectrum. And his model did work. With each jump from one orbit to another corresponding to a specific energy, and therefore to a specific wavelength of light, it explained the spectrum of light emitted by the simplest atom, hydrogen. But Planck's constant is tiny. In the same system of units where masses are measured in grams, the value of Planck's constant is 6.55×10^{-27}. Quantum effects only become important for particles which have masses, in grams, roughly equal to (or less than) this size. The mass of an electron is 9×10^{-28} grams, so quantum effects are extremely important for electrons. They are increasingly less important for bigger things, so that anything much bigger than an atom is not profoundly affected by quantum processes (except, of course, in the sense that it is made up of atoms, which are affected by quantum processes).

It is hard to get a feel for just how small the quantum world is – in a sense, how far away from us it is. But try this. If an object were 10^{-27} cm across, it would take 10^{27} such objects to stretch across a single centimetre, which is about the size of a sugar cube. That number, 10^{27}, is about 10^{16} times the number of bright stars in the Milky Way Galaxy. It is the number you would get if you took the number of stars in the Galaxy (10^{11}), multiplied it by itself (giving 10^{22}), and *then* multiplied the square by 100,000. If you took 10^{27} sugar cubes and laid them in a line, touching each other, they would span a distance of a billion light years, about a tenth the size of the known Universe. In round terms, the scale on which quantum processes dominate is as much smaller than a sugar cube as a sugar cube is smaller than the Universe. On this sort of approximation a few powers of ten don't count for much, and we can say that both a sugar cube and a human being come out roughly in the middle of the range of sizes, halfway between the quantum scale and the Universal scale, when measured in this logarithmic ('powers of ten') fashion.

So, although Bohr's model worked (especially as an explanation of the origin of spectral lines), it should be no surprise to learn that this first attempt at tackling the properties of the quantum world soon needed improvement, as new discoveries were made. The most important of those discoveries concerned the nature of electrons, which, as we have just explained, are the archetypal quantum entities, having just the right mass to be profoundly affected by processes in which Planck's constant has to be taken into account.

During the first quarter of the twentieth century several puzzling discoveries about the behaviour of electrons emerged. They can be summed up in terms of the behaviour of electrons in atoms. The Rutherford–Bohr model of the atom as like a miniature Solar System, with the electrons like tiny little balls whizzing around like planets, worked very well at explaining spectroscopy (provided the quantum behaviour of electrons jumping around energy levels was added in). But atoms themselves still behaved like smooth, hard spheres when, for example, they worked together to produce the pressure a gas exerts on the wall of a box. It is hard to see how a few electrons orbiting around the nucleus could combine to give the appearance of a smooth, hard surface to the atom.

Turning the analogy around, if a star with another planetary system around it approached our own Solar System, it would not bounce away when the outermost planets in the two systems came into contact. The visiting star would plough right through the Solar System, hurling planets out of the way, and, if it were on the right course, collide with the Sun itself. In the same way, you can't imagine one or two, or even a dozen, little electrons orbiting the nucleus of an atom providing an effective barrier when atoms collide. It might work for uranium, with 92 electrons whizzing around. But even hydrogen behaves as if the nucleus is completely surrounded by its single electron, with a uniform distribution of negative electric charge doing its best to conceal the positive charge in the nucleus. How could this be?

The answer began to emerge in the early 1920s, when the French physicist Louis de Broglie realised that the equation that uses Planck's constant to relate the wavelength of light to an equivalent energy for a particle (the photon) could be turned around. Essentially the same equation could be used to take the energy of a particle (such as an electron) and give you an equivalent wavelength for the appropriate quantum entity. De Broglie suggested, in his PhD thesis, that electrons might have to be treated as waves, under some circumstances – just as light had to be treated as a stream of particles under some circumstances.

De Broglie's thesis supervisor, Paul Langevin, couldn't decide if this was a work of genius or complete rubbish, so he sent the thesis to Albert Einstein to assess. It struck a chord. Einstein had just been involved in the definitive work which had established that light really does have to be treated in terms of photons, and that the reason why atoms can only emit or absorb light in certain quantities is because light really does only exist in packets of a certain size – quanta. It is quantised, in a manner analogous to the way cash is quantised in pennies, and an electron in an atom can only emit or absorb one photon at a time. So Einstein was already working towards the idea that all quantum entities might be both particle and wave, and he replied to Langevin that de Broglie's thesis had to be taken seriously.

De Broglie's idea, combined with the growing evidence for the reality of photons, led physicists in the second half of the 1920s to develop a complete theory of quantum physics (often known as quantum mechanics), based on the idea of wave–particle duality – that *all* quantum entities share the same schizophrenia that light has. In principle, this duality applies to everything – there is a waviness, for example, associated with the desk I am sitting at, or with my own body. But because the weight (strictly speaking, the mass) of the desk, or my body, in grams is vastly greater than 6×10^{-27}, the corresponding wavelength is vanishingly small, and can be ignored. I don't see any fuzzy edges to my desk caused by its waviness. It is only for things like electrons and photons that the two facets of reality have equal importance.

Among other things, this resolves the puzzle about how an atom that contains only a few electrons can show a smooth face to the world and behave like a little, hard sphere. Every electron in the cloud around the nucleus has to be thought of not as a tiny hard ball, but as a wave, spread out around the entire nucleus. The single electron in a hydrogen atom really does form a spherical cloud around the nucleus, all by itself. Electrons in higher energy levels can form clouds with more complicated shapes, some of them in dumb-bell-like lobes on either side of the nucleus. But the overall effect is always that the nucleus is completely surrounded by its electron cloud.

The electron clouds corresponding to different energy levels in the same atom can interpenetrate to some extent, so the best picture of the atom is not like a Solar System with electrons in individual orbits, widely spaced out from the nucleus, but something more like a series of onion skins, clouds of electric charge completely surrounding the nucleus, and quite close to one another. The change in energy levels when an atom absorbs or emits a photon can then be thought of not so much as like a ball bouncing up and down a staircase, but as like the different notes that can be played on a single guitar string by making it vibrate at different harmonics.

In 1927, the new quantum physics was vindicated when two teams of researchers, one of them headed by George Thomson (the son of J. J.) carried out experiments which unambiguously showed electrons behaving like waves when they bounced off the atoms in a crystal. In 1906, J. J. Thomson had received the

Nobel Prize for discovering the electron and for proving that it is a particle. In 1937, George Thomson received a share of the Nobel Prize for proving that electrons are waves. Both of them were right, and both prizes were merited. Nothing demonstrates better the weirdness of the quantum world and the need to remember that models are only aids to the imagination, not the ultimate truth. The neatest way to think about the effects of this wave–particle duality is to say that a quantum entity such as a photon or an electron travels as a wave, but arrives as a particle. Left to their own devices, things like electrons spread out like waves; but when you measure their position, they are instantaneously concentrated at a definite location – a phenomenon known as 'the collapse of the wave function'. Then they start spreading out again until they interact with something.

The new quantum physics, built upon the idea of wave–particle duality, has at its heart one of the most famous (but not always one of the best understood) concepts in physics – the uncertainty principle, discovered by Werner Heisenberg at the end of 1926.

It is a feature of waves that they are spread out things, and that they never can concentrate literally 'at a point'. The same is true of electrons. They may be located quite tightly – for example, when one of the electrons in the beam that paints the picture on your TV screen hits the screen and makes a tiny splash of light – but never at what a mathematician means by a point, a dot with zero dimensions. The waviness of an electron means that there is always some uncertainty about its position, and the same applies to all quantum entities. Heisenberg discovered a mathematical relationship (once again related to Planck's constant) that expresses this uncertainty. It turns out that for a quantum entity such as an electron, the more its position is constrained (the tighter its wave is squeezed), the less certainty there is about where it is going next (in effect, its velocity). You can picture this as a bit like squeezing a strong spring between your finger and thumb: it will ping out in some unpredictable direction. Alternatively, the more precisely the velocity of a quantum entity is determined, the less certainty there is about its position – the more 'wavy' it becomes. (The spring leaping out of your grip moves in a definite direction, but it has expanded and is stretched out to its maximum length, not concentrated at a point.)

What is extremely important about all this is that this is not due to some deficiency in our measuring apparatus – it is not that human beings are too clumsy to be capable of measuring accurately the position and the velocity of an electron at the same time. Uncertainty is a property built into the very nature of quantum objects. An electron does not have a precisely defined position and a precisely defined velocity at the same time; it cannot, in itself, 'know', simultaneously, both where it is and where it is going.

This has intriguing implications in the world of particle physics, which we discuss in the next chapter. But in the world of atoms it has one implication of supreme importance, which fits the last piece into the jigsaw puzzle that is the Rutherford–Bohr model of the atom. Quantum uncertainty says that the 'wave packet' describing an electron must always be bigger than a certain size (rather like the way there is a limit to how much you can compress a spring). It turns out that, under everyday conditions, this is bigger than the size of the nucleus of an atom. That is why none of the electrons in an atom can fall into the nucleus – there is no room there for their wave function to fit. Two electrons can fit into the energy level closest to the atom, eight into the next energy level outwards, eight more in the next level, and so on (this has implications for chemistry, which we discuss in chapter four).

At the other boundary of the atom, on the outside, the behaviour of electrons that we have just described neatly explains those features of atoms that Feynman thought were so important, and so revealing. The nucleus is a ball of positive electric charge, and it is surrounded by a cloud of negative electric charge which, overall, exactly balances and cancels out the positive charge on the nucleus.

But imagine what it looks like (or feels like) to another identical atom on one side of the first atom. The electron cloud on the outside of the second atom can only see half of the electron cloud on the first atom, shielding half of the positive charge inside. So the cloud feels half of the charge on the nucleus of the first atom tugging it (and therefore the whole of the second atom) that way. Similarly, the electron cloud in the first atom feels an attraction from the partially shielded nucleus of the second atom (atoms attract one another when they are a little distance apart). But when the two atoms get so close that their electron clouds are, in effect, touching one another, the negative charge on those clouds, the same kind of charge on each cloud, will repel them from each other (atoms repel upon being squeezed into one another). But in all of this, the two nuclei do not notice one another, because they are shielded from each other by a double layer of electrons.

In terms of everyday life, what goes on deep inside the atom, within the nucleus itself, is not of any practical importance. The way that the attraction between atoms works, and how the arrangement of electrons in atoms encourages some atoms to join together to form molecules, is of key interest and importance to everyday life, and explains how we got to be here in the first place, as we shall see later in the book. Indeed, if you are not particularly interested in how things work at an even deeper level, you already know enough to skip on to chapter four and find out how atoms join together to make molecules, including the molecules of life. But having got down to the level of things within the atom, and introduced the ideas of quantum physics, it would be a shame to leave without giving you a quick glimpse of what goes on at the deepest levels yet probed by physicists, where particles, fields, and quantum effects rule.

CHAPTER THREE
PARTICLES AND FIELDS

When physicists try to describe the way the world works at the most fundamental level, the most important component of their toolkit is the concept of a field. This is also one of the most familiar ideas in physics from our schooldays, when we all saw how the field of force around an ordinary bar magnet can be made visible, by holding the magnet directly under a sheet of paper which has been sprinkled with iron filings, or some other magnetic material. As the sheet of paper is gently tapped, the iron filings line up in arcs linking the north and south poles of the magnet. And it was, indeed, the study of magnetism (and electricity) that led the nineteenth-century pioneer Michael Faraday to introduce the idea of fields into physics.

Faraday's story reads like fiction. He was the son of a poor blacksmith and received only the most basic education, but became apprenticed to a bookbinder in London, and got hooked on science by reading the entry on chemistry in a volume of the *Encyclopaedia Britannica* which he was rebinding for a customer. His sterling efforts at self-education were rewarded in 1813, when Faraday (then aged twenty-one) obtained a post as assistant to Humphry Davy at the then new Royal Institution. This was the lowest possible rung on the scientific ladder, and literally involved, among other things, duties as a bottle-washer. But Faraday rose to succeed Davy as head of the Royal Institution, to turn down the offer of a knighthood from Queen Victoria, and twice turn down the offer of the Presidency of the Royal Society. He was a modest man, a member of a strict religious sect, and did not approve of personal honours of this kind.

Although he started out as a chemist, Faraday became intrigued by the nature of electricity and magnetism, which began to be recognised as different aspects of a single phenomenon at the beginning of the 1820s. The key experiment was carried out by Hans Christian Oersted, in Denmark, who found that when a magnetic compass is held near a wire that carries an electric current, the needle of the compass (which is a tiny bar magnet) is always deflected to point at right angles to the wire. So an electric current produces a magnetic force that influences the compass needle.

Progress was slow, and hampered by several false starts. But it was Faraday who established, by the end of 1831, the nature of the link between electricity and magnetism. An electric current flowing in a wire is caused by electric charge (we now know, electrons themselves) moving along the wire. A *moving* electric charge always creates a magnetic force (which, in this case, deflects the compass needle). Faraday found that in a similar way, when a magnet is moved past a wire, or pushed into the mouth of a coil of wire, while the magnet is moving it creates an electric current in the wire. A *moving* magnet always creates an electric force (which, in this case, makes the electrons in the wire move).

These two discoveries formed the basis of the electricity generator, or dynamo (in which electricity is produced by rotating magnets that whiz past coils of wire) and the electric motor (in which an electric current coursing through coils of wire makes a set of magnets attached to the drive shaft of the motor rotate).

Shortly after Faraday had discovered the dynamo effect, in 1831, the Prime Minister, Robert Peel, visited the Royal Institution and saw a demonstration of the phenomenon. In one of the most widely quoted exchanges in the history of science, he asked Faraday what use the discovery was. Faraday replied, 'I know not, but I wager that one day your government will tax it.'

The practical importance of Faraday's work does not need spelling out here, and it has, indeed, provided a rich source of revenue for governments. Within fifty years of him making that remark, electric trains were running in Germany, Britain and the United States. If he had done nothing else, Faraday (who was forty in 1831) would have been remembered as one of the greatest scientists of the nineteenth century. But after 1831, rather than developing the practical applications of the discoveries (which he left to others), Faraday became deeply interested in the nature of the link between electricity and magnetism, and how these forces could reach out across space to influence objects with which they had no direct contact. And, from the outset, he appreciated the similarity of this reaching out to the way gravity works, with the Sun reaching out across space to hold the planets in their orbits.

In the experiment with iron filings scattered on a sheet of paper, the lines traced out by the filings show the path that would be followed by a tiny magnetic pole, if it were free to drift along a trajectory from one pole of the large magnet to the other. This led Faraday to start thinking in terms of 'lines of force' (a term that he invented in 1831), stretching out from each pole of a magnet and, similarly, from each electrically charged particle. On this model, if a wire moves relative to a magnet it cuts through the lines of force, and it is this cutting that causes the electric current to flow.

Before this time, forces like gravity were imagined to operate as 'action at a distance'. The force the Sun exerted on the Earth and the other planets was not regarded as taking any time to reach out through space, but as acting instantaneously, at a distance, to keep a grip on the planets. But Faraday thought about how the lines of force would spread out, and build up what came to be called a field of force, when a current began to flow in a coil of wire.

Clearly, he reasoned, this would take time. A suitably arranged coil of wire carrying an electric current produces a magnetic field that is indistinguishable, at a distance, from the field of a bar magnet. But when there is no current flowing in the coil, there is no magnetic field. Faraday was convinced that when the current in the coil was switched on, the magnetic field would spread outwards from it, so that compass needles near the coil would be deflected first, and compass needles further away later, as the magnetic field spread outwards. We now know that this influence spreads at the speed of light, far too fast for the delay to be measured in any nineteenth-century laboratory experiments; but Faraday was thinking along the right lines.

The idea was so outrageous at the time, though, that he did not go public at first. Instead, in 1832 he left a sealed note in a safe at the Royal Society, to be opened after his death. Among other things, the note spelled out Faraday's idea that,

> when a magnet acts upon a distant magnet or piece of iron, the influencing cause (which I may for the moment call magnetism) proceeds gradually from the magnetic bodies and requires time for its transmission ... I am inclined to compare the diffusion of magnetic forces from a magnetic pole, to the vibrations upon the surface of disturbed water, or those of air in the phenomena of sound: i.e. I am inclined to think the vibratory theory will apply to these phenomena, as it does to sound, and most probably light.

Twelve years later Faraday did go public with these ideas, which paved the way for James Clerk Maxwell's work on the nature of electromagnetism and light in the 1860s.

Maxwell found a set of equations that describes how electric and magnetic fields interact with one another. There are just four of these equations, and together they tell you everything you could ever need to know about electromagnetism, provided you do not venture into the realms of quantum physics. Every phenomenon involving electricity and magnetism at the classical level can be solved using Maxwell's equations; and they came with a built-in bonus.

Among other things, Maxwell's equations explain how an electromagnetic wave moves. Think of an electric wave which moves off through space, energised, perhaps, by an electron jiggling up and down in a wire. As the electric wave moves up and down, it is varying. So it must make a magnetic wave, which moves alongside the electric wave. But the magnetic wave is also varying, because the electric wave varies. So the magnetic wave must create an electric wave, which moves alongside the magnetic wave. The overall effect, described with precision by Maxwell's equations, is of a combined electromagnetic wave moving through space, with the electric and magnetic components in step with one another. The bonus in Maxwell's equations is a number that specifies the speed with which the wave moves.

This number can be worked out from experiments measuring the properties of isolated electric and magnetic fields. And the number that comes out of the experiments is the speed of light, even though, as Maxwell put it, 'the only use made of light in the experiment was to see the instruments'.

Maxwell discovered that electromagnetic waves move at the speed of light, and realised that light must be a form of electromagnetic wave. This conclusion became inescapable when radio waves, predicted by Maxwell and also moving at the speed of light, were discovered by Heinrich Hertz in the 1880s. This is why physicists were so surprised, in the early part of the twentieth century, when they found that in the quantum world light sometimes has to be treated as a stream of tiny particles. But maybe Faraday wouldn't have been so surprised.

Faraday first expressed his ideas about lines of force and fields in public in two lectures that he gave at the Royal Institution, in 1844 and 1846. His ideas were way ahead of their time – in those days it was still widely accepted that 'empty space' was filled with a mysterious substance called the aether, which carried the ripples of light waves in the same way that ripples in a pond are carried by the water in the pond. In his 1846 lecture, spelling out more or less the model we have just outlined, but without Maxwell's mathematics, Faraday specifically said that his aim was 'to dismiss the aether, but not the vibrations', suggesting instead that the vibrations were associated with the lines of force, like twanging guitar strings. He pointed out that the propagation of light takes time, which fits in with the idea that a ripple is moving along a line of force, and he surmised that gravity must operate in a similar way. But in some ways what he said in the earlier lecture, on 19 January 1844, is even more fascinating.

As well as dismissing the concept of the aether, Faraday wanted to do away with the idea of atoms – this before the idea of atoms had been completely assimilated into science. From the perspective of the success of the model we described in the last chapter, this sounds crazy. But from the perspective of the quantum field theory of the twentieth century, it would make perfect sense, if we replace the nineteenth-century meaning of the term 'atom' with the equivalent modern term, 'particle' –

another example of the need to use different models when describing the same thing under different sets of circumstances.

Faraday argued that there could be no real difference between so-called space and the atoms ('particles') in space. He said that atoms (particles) should be regarded merely as concentrations of forces, and that instead of thinking of an atom (particle) as the origin of a web of forces (the field) stretching out around it, what we should think of as the fundamental reality was the field of force itself, with the particles treated simply as concentrations in the lines of force – knots in the web.

In order to make his point, Faraday asked his audience to imagine a 'thought experiment' in which the Sun was sitting on its own in space. What would happen to the Earth if it were suddenly, by magic, placed in position at its appropriate distance from the Sun? Would some force have to travel outwards from the Sun and grab hold of the Earth before it could begin to feel the Sun's gravity? Or would it feel the Sun's gravitational pull immediately?

Faraday said that even if the Sun were alone in space, its gravitational influence would spread everywhere, including through the place where the Earth was about to appear. The Earth would not be dropped into 'empty space' but into a web of forces (the field), and it would respond instantly to the nature of the field at the position of the Earth itself. What matters to the Earth is the nature of the field at the location of the Earth, not the nature of the source of the field (in this case, the Sun) – if, indeed, the idea of a source for the field was correct. To Faraday the field was the reality, and matter was just a region where field was concentrated, or knotted up. In the course of those two lectures, he dismissed both the idea of an aether and the idea of materially real particles, leaving a picture of the Universe as nothing more (or less) than a web of interacting fields, knotted here and there. This is almost exactly the way a modern quantum field theorist would describe the Universe.

The classic example, of course, is the photon itself. A photon can be regarded as a knot in the electromagnetic field, a little tangle of electromagnetic waves. But remember that physicists discovered in the twentieth century that even what we are used to thinking of as material particles, things like electrons, also have a wave aspect to their nature. In quantum field theory each kind of particle has its own field – so there is, for example, an electron field, which fills the Universe. What we perceive as the particles (electrons, or whatever) are knots in the appropriate field, in the same way that photons are knots in the electromagnetic field.

It is strange to think of there being an electron field that fills the Universe, and which we do not, in the ordinary way of things, notice. It is even stranger to think of there being a different kind of field for each kind of particle, as well as the more familiar electric and magnetic fields, and so on. It sounds like the Universe should be jam-packed with a confusing tangle of fields (which, on this model, it is), and that we ought to be aware of this (which we are not). But remember that you do not notice the electromagnetic fields that permeate the environment – at this moment, there are signals from dozens of radio and TV stations going right through you.

The situation with all these matter fields, as they are known, is rather like the way we perceive – or rather, don't perceive – the atmosphere of the Earth. On a still day, you just don't notice that you are immersed in a sea of air. Because it is the same everywhere, there is no way to tell that it is there at all. The only reason you notice the air is when it is in motion, and the motion is caused by irregularities, ranging from little eddies to great storm systems and whirlwinds. The material particles that we notice – the electrons and so on – are the equivalent of eddies in the matter fields and, like eddies, their influence extends a little way out into their surroundings. We only perceive the irregularities, not the smooth, bland uniformity of the fields that fill the rest of the Universe.

This turns out to be a very useful model when probing the deeper layers of the quantum world; but for our purposes it is enough to know that this is the way quantum field theorists view the world, and to be impressed by Faraday's insight. In terms of getting an outline picture of what goes on deep within atoms, we can still work with a model that is a mixture of particles and fields, and explain the behaviour of these entities by analogy with the way electrons interact with electromagnetic fields, and with each other. The idea of particles is still useful, whether or not you think of them as knots in a field.

When two electrons come near to each other they are repelled from one another, because they each have the same (in this case, negative) electric charge. But how does the repulsion work? By sticking with the image of electrons as particles, and applying quantum theory to the electromagnetic fields that surround these particles, we can get a clear image of what is going on. The interaction between the two electrons involves a stream of photons (carriers of the electromagnetic force) which travel from one electron to the other (and, indeed, from the second electron to the first). Think of these as like a stream of machine-gun bullets. Each electron emits a stream of particles and recoils, while at the same time each electron is hit by a stream of particles and knocked away. No wonder the two electrons repel each other.

It is harder to envisage why oppositely charged particles (for example, an electron and a proton) should attract one another, but that is just the way it happens. One analogy, which helps a bit, is to think of a group of athletes involved in a training exercise where they are jogging along while passing a heavy medicine ball from one to another. They have to keep close together, because they can't throw the medicine ball very far. But sometimes this attraction works, in the particle world, even when very light particles are being exchanged. A stream of photons from an electron, arriving at a proton, doesn't push it away, but instead pulls it towards the electron, and vice versa.

All of this can take place at a considerable distance, because photons are easy to make. They don't have any mass at all, so although they do carry some energy, there is no mass energy involved, and charged particles can make streams of photons without losing much energy. Once made the photons can travel, in principle, for ever across space. So electromagnetism is a very long-range force and can, in principle, reach out right across the Universe (just as gravity does, and for the same reason – the graviton also has zero mass). In fact, the range of electric and magnetic forces is limited by their tendency to be balanced out on a smaller scale. Because there is one electron for every proton in an atom, there is no overall electric charge to influence the world at large.

Things are significantly different when we get down to the scale of the atomic nucleus. There, the protons and neutrons are held together by another force, the strong nuclear force, which prevents the concentration of all the positive charge in the nucleus from blowing it apart. But this force is not directly connected with neutrons and protons, but with a deeper layer of structure within them, at the level of quarks.

There is all kinds of evidence for the existence of quarks, but the best, and most direct, comes from experiments carried out at the end of the 1960s and in the 1970s, when very energetic beams of electrons were fired at atomic nuclei, and the way the electrons bounced off (scattered from) the particles in the nuclei was studied. In a manner strikingly reminiscent of the way Ernest Rutherford's team had found structure inside the atom by scattering alpha particles from the nucleus, these experiments, pioneered by researchers at Stanford University in California, showed structure inside protons and neutrons by scattering electrons from the particles there.

Protons and neutrons are made up of quarks – in each case, three quarks. But only two different types of quark are needed to explain the structure of protons and neutrons. These have been given arbitrary names, which are only meant as labels – one kind of quark is called 'up' and the other is labelled 'down'. But this does not mean, in any sense, that they point in different directions, or sit at different heights – the labels could just as well have been Gert and Daisy.

Each up quark carries an amount of electric charge that is positive and equal in size to two-thirds of the charge on an electron. Each down quark carries an amount of electric charge that is negative, and equal in size to one-third of the charge on an electron. A proton is composed of two up quarks and one down quark, which together give it 1 unit of positive charge ($\frac{2}{3} + \frac{2}{3} - \frac{1}{3} = 1$), while each neutron is made up of two down quarks and one up quark, and therefore has no overall electric charge ($\frac{2}{3} - \frac{1}{3} - \frac{1}{3} = 0$).

The most intriguing feature of quarks is that they are never seen in isolation, only in triplets or in pairs. The pairs of quarks (always, in fact, a quark-antiquark pair[1]) are known as mesons, and it is mesons that hold atomic nuclei together, constantly being exchanged among the protons and neutrons that make up the nucleus. Mesons carry a force between the components of the nucleus (the nucleons) in a way analogous to the way photons carry the electromagnetic force. But, unlike photons, mesons have mass. So it takes a lot of energy to make a meson. The only way that mesons can be made inside a nucleus is because quantum uncertainty comes into play.

This time the relevant uncertainty involves the amount of energy that there is associated with each of the nucleons. Just as the position and momentum of a quantum entity cannot be determined precisely, so the amount of energy locked up in a quantum entity cannot be determined precisely – not because of the imprecision of our measuring apparatus, but because the Universe itself does not 'know' precisely how much energy is sitting there at any instant. For a short enough time (a time related to Planck's constant, so therefore very short indeed), energy can appear out of nothing at all, provided it disappears again within its allotted time. The more energy that appears, the shorter the time it can stay. But if you have enough energy, it can temporarily turn itself into particles during its brief lifetime.

This is where the mesons that hold the nucleus of an atom together come from. They appear out of nothing at all, as so-called vacuum fluctuations of the quantum fields. When this happens, the quantum rules say that each particle that is produced must be accompanied by its antiparticle counterpart. Instead of isolated quarks appearing out of the vacuum, you get quark–antiquark pairs. In other words, mesons. But they can only live for a very short time, the time allowed by quantum uncertainty, before giving their borrowed energy back to the vacuum and disappearing once again. They live just long enough to be exchanged with the nucleon next door, holding the nucleus

[1] An antiparticle is a kind of mirror image of a particle, with opposite properties, such as charge. So an anti-electron, for example (also known as a positron) has one unit of positive charge, but the same mass as an electron.

together (this is where that medicine-ball analogy may be helpful). But the range of the strong force that results is so limited that nothing outside the nucleus, not even the nucleus of the atom next door, can ever feel it.

But why do quarks only come in triplets or in pairs? Because they are held together, in their turn, by the exchange of yet another kind of quantum field particle. When it came to naming these particles physicists allowed themselves a little joke – the particles are called gluons, because they glue quarks to one another. Gluons are the quanta associated with yet another field, and operate in almost the same way as all the other forces we have discussed. But there is one crucial difference. The glue force between two quarks gets *stronger* when the quarks are further apart.

Instead of thinking of the exchange of gluons between two quarks as like a stream of particles knocking them apart, think of it as like a very powerful piece of elastic joining the two quarks together. When the quarks are close together the elastic is loose and floppy and the quarks jostle about. But when the quarks try to move apart (even as far apart as from one side of a proton to the other), the 'elastic' stretches and pulls them back together. The more they move apart, the more the elastic stretches, and the more strongly they are pulled back together.

The only way that a quark can be knocked out of a nucleon or a meson is if something hits it so hard – with so much energy – that it 'breaks the elastic'. This can, in fact, happen when an electron moving at nearly the speed of light smashes into a proton in the kind of experiments we just described. But it can only happen if the energy provided by the impact of the electron is great enough to be turned into the mass of *two* new quarks (strictly speaking, a quark–antiquark pair). Where the 'elastic' breaks, each broken end attaches to one of the two new quarks. One end snaps back into the proton, restoring it to normality. The other end links the new antiquark to the escaping quark – instead of a single quark escaping from the site of the impact, we see a meson. We never, under any circumstances, see a lone quark.

So the glue force is the real force that holds the components of nuclei together, including the quark–antiquark pairs that make up mesons. The strong nuclear force that operates between nucleons is really a relatively weak vestige of the glue force, and they are not regarded as two genuinely distinct forces of nature. Along with electromagnetism and gravity, that means we have so far mentioned three genuinely different kinds of force operating in the particle world. There is just one more to mention, the so-called weak force (because it is weaker than the strong force). This is the least like our everyday understanding of what is meant by the term 'force', and in recognition of this physicists often refer to the four forces of nature as the four 'interactions'.

The weak interaction doesn't so much hold things together as break them apart. In particular, it interacts with neutrons, converting them into protons. This is called beta decay, and it involves the neutron emitting an electron and a particle called a neutrino (strictly speaking, an anti-neutrino). More accurately, the neutron interacts with the field associated with the weak interaction and emits one of the quanta of that field. These field quanta are called intermediate vector bosons, and they have mass so that their range (the range of the weak interaction) is limited by the rules of quantum uncertainty. The boson then turns into an electron and an anti-neutrino. (There are other ways in which the weak field can interact with other particles, but we don't need to discuss them here).

The pleasing thing about all this is that it leaves physicists with a very neat, fundamental symmetry in the particle world. There are four kinds of fundamental forces, and also four kinds of fundamental particles (the up and down quarks, the electron, and the neutrino). This is absolutely everything you need to explain everything that you can see in the Universe. Unfortunately, for reasons that nobody has fathomed, some sort of duplication (triplication, even) seems to be at work in the laws of physics. In experiments involving high-energy collisions between beams of particles, physicists find two more families of particles, with successively higher masses. These are not particles that were in any sense 'inside' the particles that were smashed together in these experiments, in the way that quarks are inside nucleons. Instead, they are particles that have been created out of pure energy (like the quark–antiquark pairs at the place where the gluon elastic breaks), and in extreme cases they include varieties of particles that may not have existed naturally in the Universe since the Big Bang. There is a heavier counterpart to the electron, which is associated with its own kind of neutrino, and heavier counterparts to both the up quark and the down quark. And *then*, as if that were not enough, there is a heavier set still – of everything.

None of these heavier versions of the basic particles play any role in the way the Universe works today (although they may have been important in the Big Bang; *see* chapter twelve). They are all unstable, and decay, eventually turning into their lighter counterparts, as they release energy. They do provide particle physicists with plenty of interesting things to investigate, and scope to develop plenty of theories about the reasons for their behaviour. But in this book we are going to concentrate on the way the everyday Universe is, not on the more exotic possibilities of how it might have been, and we shall say no more about these heavier versions of the fundamental particles for now. Instead, before coming back from the depths of sub-nuclear physics to the world of molecules, we want to take stock of those four fundamental forces of nature, and how physicists are trying to develop an overall picture – a unified theory – to explain the behaviour of those forces (or fields), and of the basic set of four particles, in one package.

Gravity is the weakest of the four forces, but it is the most obvious in everyday life, and was the first to be studied scientifically, because of the way it adds up, and because of its very long range. In any lump of matter the gravity of all the individual particles in the lump adds together. There is no cancelling out in the way that the positive and negative electric charges in an atom cancel each other out. The range of gravity, like the range of electromagnetism, is, in principle, infinite, because the relevant field quanta (gravitons) have zero mass, like photons. But that doesn't mean the force from a lump of matter is the same strength everywhere. In fact, the gravitational force from an object drops off as one divided by the square of the distance from the object (the famous 'inverse square law'); so if you are twice as far away the force is one quarter as strong, three times further away the force is one ninth as strong, and so on.

Electromagnetism is much stronger than gravity, but this is not obvious in the everyday world because of the way both electricity and magnetism come in two varieties – positive and negative charge, north and south poles – which cancel each other out. But, like gravity, both electricity and magnetism obey inverse square laws, wherever they are not completely cancelled out. As Maxwell's equations showed, they are different aspects of a single force, both carried, in the quantum field model, by massless photons.

The best insight into the relative strengths of gravity and electromagnetism comes from thinking about what happens when an apple falls from a tree. The force that is trying to hold the apple on to the branch where it grew is the electromagnetic force, operating between a few atoms in the stalk holding the apple to the branch. The force pulling the apple down is the force of gravity exerted by every atom in Planet Earth pulling together, a mass of just under 6×10^{24} kilograms. The gravity of all that concentration of mass, working together, is just able to break the electromagnetic bonding between a few atoms in the stalk of an apple.

The relative strengths of all four of the forces of nature can be expressed in terms of the strength of the strong force. If you set the strength of the strong force as 1, in very round terms the strength of the electromagnetic force is about 10^{-2} (one per cent as strong as the strong force), the strength of the weak force is 10^{-13} (one hundred-billionth of one per cent of the strength of the strong force) and the strength of gravity is a mere 10^{-38}. The strong force is one hundred billion billion billion billion times stronger than gravity. Even the weak force is ten million billion billion times stronger than gravity. But remember that the weak force and the strong force only have very limited ranges, operating on a scale much smaller than the size of an atom. The Universe would be a very different place if they had the same range as gravity and electromagnetism.

But that is exactly what physicists believe did happen, in the Big Bang, when the Universe was very young. We will have more to say about this in chapter twelve, but it ties in with the way physicists are trying to find a way to combine all of the forces in one mathematical package, a unified field theory.

In a sense, they are already at least halfway there. Electromagnetism and the weak interaction have already been combined in a single package, which is called the electroweak theory. It was developed in the 1960s, and is the role model for almost all subsequent attempts to develop a unified field theory.

The basis of the electroweak theory is the similarity of the roles played by the field quanta in electromagnetism (photons) and in the weak interaction (the intermediate vector bosons). Electromagnetism is a particularly simple interaction to describe mathematically, because there is only one kind of field quantum, the photon, and it has no charge or mass. In the weak interaction, the intermediate vector bosons come in three varieties. One has positive charge, one negative charge, and the other zero charge; and they all have mass. Otherwise, though, they are just like photons. If they were not hampered by the limit on their range imposed (because of their mass) by quantum uncertainty, they would behave exactly as photons do. If the Universe were hot enough, however, the intermediate vector bosons would not have to rely on quantum uncertainty for their existence. The background energy of the Universe would be enough to make real intermediate vector bosons, out of the hot radiation filling the Universe, which could live for ever and have infinite range, just like photons.

The electroweak theory describes how particles would have interacted under those conditions, with photons and intermediate vector bosons on an equal footing. Crucially, it also describes how the different forces would have split apart when the Universe cooled and expanded away from the Big Bang, and it predicts that this splitting would produce just the differences between the two forces that we see today – provided that the intermediate vector bosons have a certain mass.

The intermediate vector bosons were manufactured out of pure energy in particle collision experiments at CERN, near Geneva, in the early 1980s, and they were found to have exactly the masses predicted by the electroweak theory.

The next stage will be to find the same kind of theory which includes the strong force with the electroweak interaction. This means bringing gluons in on an equal footing with photons and intermediate vector bosons. In principle, the same approach ought to work. But because gluons are even more massive than intermediate vector bosons it only works at even higher temperatures, corresponding to even earlier times in the Big Bang.

The way you test these ideas, of course, is by smashing particles together in accelerators, and thereby briefly re-creating the kinds of conditions in which the fundamental forces are on an equal footing. Such experiments are, as we have mentioned, powerful enough to test the electroweak theory quite fully, and make sure that the theory describes accurately the way real particles behave. Accelerators powerful enough to test the predictions of the equivalent theories that bring in the strong force as well (usually known as Grand Unified Theories, even though they do not include gravity) cannot be built. It is not too difficult to come up with theories of this kind, but a theory untested by experiment is useless – 'if it disagrees with experiment it is wrong'. This is particularly important in this case, because there are also more complications to test than in the electroweak theory, because there are eight different kinds of gluon to take into account.

The variety of particles involved allows physicists to construct different kinds of Grand Unified Theory, each making different predictions about the way matter should behave at very high energies. But in order to test these theories and find out which one (if any) is right, we would need an accelerator a thousand billion (10^{12}) times more powerful than the accelerators which were used in the 1980s to test the electroweak theory. An accelerator as big as the Solar System would still be too small to do the job. This is why, as we shall see in chapter twelve, particle physicists are focusing much of their attention on cosmology, using their theories to predict what happened in the Big Bang, and then comparing their predictions with the way the Universe actually looks.

And even then, we have not included gravity in the package. The Holy Grail of theoretical physics is a kind of Super Grand Unified Theory, which would bring gravity in as well.

Unfortunately, even though gravitons have zero mass this is not easy; you still have to go to even higher energies to make gravity as strong as the other forces. Because gravity is so weak, the step up in energy involved to make the same approach work here is so enormous that it is hard to see how it will ever be made, and such progress as there is comes from a completely different direction – a theory which replaces the familiar idea of particles as points at which the field quanta are located with the idea of little loops of 'string'.

These strings form loops which are much smaller than the size of a particle such as a proton, but the crucial point is that they do have a definite size; they are not mathematical points. In standard quantum field theory, the truly fundamental entities such as quarks and electrons (which seem to be fundamental today, anyway) are thought of as having zero size. The typical size of a loop of string would be very, very small – they exist down on the scale where quantum uncertainty is important, which means they have diameters of about 10^{-33} cm. A loop of string is as much smaller than an atom as an atom is smaller than the Solar System.

Even though there is no way to test the predictions of string theory directly – there are no conceivable experiments which probe structures on as small a scale as this – there are two reasons why many physicists today think that strings represent the ultimate truth, and are the basic building blocks of matter.

The first is that they get rid of unwelcome infinities that plague any theories based on older ideas that envisage particles as fundamental points. That would mean that the particles have zero volume, and as a result, somewhere along the line in the calculations you inevitably end up dividing by zero. For example, the electric force obeys an inverse square law. Its strength at a certain distance from the source is proportional to one over the square of the distance from the source. So as you approach the source of an electric field, the strength increases as the distance decreases. But if the source has zero size, the distance can decrease all the way to zero. If the source – an electron, say – has zero size, the source itself feels a force which is proportional to one divided by zero (worse – one divided by zero squared!). In other words, infinity.

There are ways to get around this, using a trick called renormalisation. In effect, this means dividing one infinity by another one to give a sensible (finite) number. At first guess you might think that infinity divided by infinity is 1, just as 2/2 is 1, or 51234/51234 is 1; but infinity is tricky stuff, and you can get just about any answer you want from this kind of division. For example, think of a number which represents the total of possible integer numbers (1 + 2 + 3 + . . .). Obviously, it is infinity. Now double every number in the sequence, and add them up again. Clearly, the answer is, once again, infinity. But how does this second infinity compare with the first infinity? You might expect that, since every number in the second infinite sequence is twice as big as its counterpart in the first infinite sequence, the infinity you get as a result of adding them all up will be twice as big as the first infinity. But think again. Your second sum only includes all the even numbers (2 + 4 + 6 + . . .). It doesn't include any of the odd numbers at all so it can only be *half* as big as the first infinity! Divide your second infinity by your first infinity and the answer you get is 0.5, not 2 (and certainly not 1).

Using this kind of trickery in the standard theory, renormalisation can be made to work, giving finite answers that match experimental measurements of things like the strength of the electric force. But it is a procedure forced upon physicists by circumstances, and one which many physicists hate. The problem of dealing with infinities in this way disappears in string theory, though, because you no longer have to deal with mathematical points that have zero size. There are none of these uncomfortable point sources (also known as singularities), and no need for renormalisation.

The second great thing about string theory is that it *predicts* gravity. As we mentioned, in the older approach to Super Grand Unified physics, it looks very difficult to find a way to bring gravity into the fold, because gravity is so much weaker than the other forces. When people started dabbling with string theory and trying to use it to provide a description of the known forces and particles, they assumed that gravity would prove equally intractable in the new theory, and didn't even try to take it into account, at first (this was in the 1970s).

The way string theory works, though, is not to have a different kind of string loop for each kind of fundamental particle or field quantum, but to regard all of the fundamental particles and field quanta as different forms of vibration on the same kind of string, rather in the way that a single violin string can play many different notes. In the mid-1970s, when a few mathematical physicists were playing with this idea as an abstract theoretical conception, they found ways to describe all of the known fundamental entities (quarks, photons, and so on) in this way. A photon would correspond to a vibration of the string loop equivalent to one note on a violin, an electron would correspond to a different vibration (equivalent to a different note on the violin string), and so on. But they also found that there is another way in which the loops of string could vibrate, which didn't match any of the particles and fields they were trying to describe.

At first they tried to find a way to get rid of the unwanted form of vibration. Then they realised that it exactly corresponded to the graviton, the quantum of the gravitational field. String theory automatically includes gravity, whether you ask for it or not!

This doesn't mean that physicists yet have their Holy Grail, their theory of everything. The mathematical complexities inherent in string theory make it difficult to develop the ideas fully, and there is still the frustration of not being able to test the ideas against experiments. Compared with the way progress in particle physics had been made previously, the discovery of string theory has happened backwards. Before, ever since the pioneering work of Rutherford, people had probed into the structure of the atom and then more fundamental particles, and developed theories to explain the results of their experiments. But this time, the theory has come along out of the pure mathematics, without any underpinning of experiment on which to build. There has never been an experiment in which particles have been made to collide together and bounce off each other in a way which automatically suggests that what is doing the bouncing is little loops of vibrating string. String theory has been memorably described as a piece of twenty-first-century physics that fell into the laps of twentieth-century physicists. But it is widely regarded as a viable alternative to the older field theory approach using the idea of point particles, and it definitely removes (or rather, never experiences) the difficulties which seem to make a complete quantum field theory of gravity an impossible dream in those theories.

However things develop over the next hundred years, though, one thing is certain. Writing at the end of the twentieth century, this is as far as we can take you in terms of probing into the innermost structure of matter. It is also true that none of these ideas, fascinating though they are, has any direct bearing on the nature of the world we inhabit. In order to have an understanding of the everyday world around us, we only need to understand the behaviour of atoms and their components, treating both protons and neutrons, as well as electrons, as fundamental particles. And we only need to worry about two of the four fundamental forces – gravity and electromagnetism.

In order to continue our description of the scientific understanding of the world we live in, it is time to turn our attention outwards, starting once again with atoms, but looking at how they are assembled into larger and larger objects. Working in this direction, the ultimate aim of science is to explain the existence of the Universe itself, and how it came to be the way it is. Somewhere in the middle range of sizes between the world of fundamental particles and the Universe at large we will find an explanation of our own existence.

When we were probing inside the atom we were looking at simpler and simpler sub-units of the physical world as we focused on smaller and smaller scales; but our first steps outward from the atomic scale to the Universe at large have the opposite effect, as we look at how the behaviour of such simple things as atoms enables them to get together to form complex things like people. The first step is to look at the way atoms latch on to one another to form molecules – chemistry.

CHAPTER FOUR
CHEMISTRY

All of chemistry can be explained by the physics of the 1930s; and a very simple model is enough to give an insight into why atoms join together to make molecules in the way that they do. We don't even need to worry about neutrinos, let alone the strong and weak forces. We certainly don't have to worry about gravity, which is far too weak to be important in chemical reactions. Also, at least in the first stages, we don't need to worry about wave–particle duality. We scarcely need to worry about the fact that the nucleus of an atom is made up of two different kinds of particle, the proton and the neutron. All we need to know to make a start on explaining chemistry is that atoms are, indeed, made up of negatively charged electrons arranged in accordance with the rules of quantum physics at some distance from the central, positively charged nucleus, and under the influence of electromagnetic forces. The basics of chemistry are incredibly simple; but the complexity of the molecules that can be built in accordance with the basic simple rules is astonishing.

The first successful explanation of the periodic table of the elements actually dates back to 1922, when Niels Bohr used a mixture of the old ideas of classical physics (including the idea of the electron as a little particle) and the new ideas of quantum physics to develop what is essentially the atomic model still taught in schools today, and which we mentioned in chapter two. This was ten years before the discovery of the neutron, which shows how little you need to know about what goes on inside the nucleus in order to understand chemistry.

The atomic number of the nucleus (which, we now know, is

equal to the number of protons in the nucleus) determines how many electrons there are in the outer layers of the atom. There has to be one electron for each proton, in order for the electric charges to balance and leave an electrically neutral atom. Whatever the atomic number, the first electron belonging to the atom goes into the lowest energy level available, which corresponds to the energy level occupied by the lone electron in an atom of hydrogen, the simplest atom. The next electron slots into the same energy level, but has opposite spin to the first electron. There are only two places available at that level. But there is more room for the electrons in the next energy level, in a sense further out from the nucleus. So the next eight electrons, corresponding to the next eight atomic numbers (that is, to the next eight elements in the periodic table), can all slot in at about the same 'distance' from the nucleus. Then the pattern repeats (with some subtleties, which are of interest chiefly to the specialists).

Each layer of electrons around the nucleus is called a 'shell'. This is a slightly unfortunate name, since it implies (incorrectly) that the outer layers do not 'see' the nucleus, but only the next shell below; but the terminology is a hangover from the early 1920s, and we are stuck with it now. One of the subtleties, which we shall not discuss in detail, is that because of the way the energy of the whole atom is altered as successive shells are filled in there comes a point where it is possible to squeeze some extra electrons into one of the *inner* shells. As a result, there is a whole string of elements (the rare earth elements) which have very similar chemical properties, because they have identical outer shells, but different atomic weights and different atomic numbers. For these elements, as the number of protons in the nucleus increases one at a time, the extra electrons are not forming new shells on the outside of the atom, but are being hidden away in the inner shells.

Bohr's explanation of the periodic table, and of the details of spectroscopy discussed in chapter two, involved filling each shell in succession for atoms with higher and higher atomic numbers (but took due account of the subtle effects which produce the rare earths). His key insight, which made chemistry once and for all a branch of physics, was the realisation that as far as its interactions with the outside world (that is, with other atoms) are concerned, just about the only thing that matters for an atom is the number of electrons in its *outermost* occupied shell. What goes on deeper inside the atom is only of secondary importance.

This is why hydrogen, which has only one electron, has similar chemical properties to lithium, which has three electrons – two of which are in the innermost shell and one of which is on its own further out from the nucleus. It is why fluorine, which has two electrons in the innermost shell and seven in the next shell, has similar chemical properties to chlorine, which has two filled shells (one containing two electrons and one containing eight electrons) and a third, outermost occupied shell containing seven more electrons. And so on. Of course, the existence of the inner shells cannot be totally ignored, and nor can the overall mass of the atom be completely neglected, which is why chlorine is not exactly the same as fluorine, and hydrogen is not exactly the same as lithium. But the similarities, in each case, are much more important than the differences.

Bohr had no idea why a shell containing eight electrons should be full (or 'closed', in his terminology) – but that doesn't matter in order to understand chemistry. What does matter is that, for whatever reasons, atoms like to have closed shells. Later, using quantum mechanics, it was possible to explain why that should be so – a full electron shell corresponds to a low energy state for the atom, and low energy states are always desirable. But we do not have to worry about that here, just as Bohr didn't have to worry about it in 1922. Bohr's insight immediately explained why certain atoms like to get together to form molecules, and (equally significantly) why other atoms are reluctant to react in this way.

The best way to explain this is by an example. Hydrogen has just one electron in its sole occupied shell, and it would like to have two there, to close the shell. Carbon has a total of six electrons, two of them in the innermost shell (which is therefore closed) and four in the second shell. It would like to have eight electrons in the outer occupied shell. One way in which it can achieve an illusion of this is by getting together with four hydrogen atoms to form a molecule of methane (CH_4).

Imagine each hydrogen atom (or rather, each hydrogen nucleus) grabbing hold, through the electromagnetic force, of one of the carbon atom's four outer electrons, while the carbon atom (nucleus) has a hold of each of the four electrons from the four hydrogen atoms. Pairs of electrons are shared between two nuclei in such a way that each hydrogen atom has a part share in two electrons, and the carbon atom has a part share in eight electrons (plus its two inner electrons). It turns out that this is, indeed, a more stable (lower energy) state than having four loose hydrogen atoms and one loose carbon atom.

The same reasoning explains why substances such as helium (with a filled outermost occupied shell of two electrons) and neon (with a filled outer shell of eight electrons, as well as a filled inner shell of two electrons) are reluctant to interact chemically at all. They have no need to – they have already achieved the chemical equivalent of nirvana.

The kind of chemical bonding that is described by this sharing of electrons between atoms is called covalent bonding. One shared pair of electrons make one covalent bond. But there is, as Bohr realised, another way in which chemical nirvana can be achieved. Think of an atom which has a single electron in its outermost occupied shell – sodium, say, which has two full shells (one of two and one of eight electrons) plus one odd electron out on its own. The easiest way it could form a closed outer shell would be to get rid of the outermost electron, if only there were somewhere for it to go.

Now think of an atom which has a single gap in its outermost occupied shell – such as chlorine, which has two full shells (one of two and one of eight electrons), plus a third shell containing seven electrons. The easiest way it could form a closed shell would be to latch on to an extra electron, if only there were one available. When sodium and chlorine get together each sodium atom can give a single electron to a chlorine atom. Both achieve chemical nirvana. But each electron that is exchanged carries with it a negative electric charge of 1 unit, so every sodium atom is left with a residual electric charge of +1 and every chlorine atom is left with a residual electric charge of –1. Since opposite electric charges attract one another, the charged atoms (usually known as ions) stick together electrically to make crystals of common salt, sodium chloride. This is called an ionic bond.

These are particularly clear-cut and straightforward examples, and in many cases the chemical bonding between atoms in a molecule involves aspects of both covalent bonding and ionic bonding. But the details need not bother us. What matters is that all chemical reactions can be explained in this way, as a sharing or swapping of electrons (or both) between atoms, in an attempt to achieve the desirable state of having filled outermost occupied shells.

Like all the best scientific models, Bohr's made a prediction that could be tested. The prediction was compared with the results of experiments, and the model passed the test with flying colours.

Even in 1922 there were still a few gaps in the periodic table of the elements, corresponding to undiscovered elements with atomic numbers 43, 61, 72, 75, 85 and 87. Bohr's model predicted the detailed properties of the elements needed to plug the gaps in the table, just as Mendeleyev had made similar predictions in the previous century. Crucially, though, Bohr's model predicted different properties for element 72 (hafnium) than rival models did. Less than a year after the prediction was made hafnium was discovered and named, and turned out to have just the properties predicted by Bohr (the other 'missing' elements were also found to have the predicted properties, although in those cases the differences between Bohr's predictions and those of other models were less clear-cut).

All of this was achieved without using the idea of electron waves, which had not been invented at the time. In a molecule such as hydrogen (H_2) each hydrogen atom was envisaged as contributing one electron to a pair, with the pair of electrons more or less between the two hydrogen nuclei so that each nucleus could get a grip on both of the electrons. But the idea of electron waves provides a way of looking at molecules which is in some ways easier to understand, once you accept that individual electrons can, indeed, be spread out over a volume of space equivalent to the size of the atom or molecule, not constrained to a single point orbiting around the nucleus. The volume occupied by a single electron in an atom or molecule is called an orbital, and in the case of the hydrogen atom this is essentially a spherical shell surrounding the nucleus. Every orbital can be occupied by a maximum of two electrons, with opposite spin to one another.

Now we have a different view of the hydrogen molecule from Bohr's 1922 model. Instead of two point-like electrons, sitting in the gap between the two nuclei, we can think of each electron as surrounding *both* nuclei.

Imagine the two nuclei sitting a little distance apart, kept away from each other by the mutual repulsion of their positive charge. The orbital that is now occupied by the two electrons can be pictured as a kind of hourglass shape (with a very fat neck) surrounding both nuclei, with one nucleus in each half of the hourglass. Each nucleus is fully surrounded by the orbital, and each feels the presence of two electrons in the appropriate orbital, closing the innermost shell. It is as if the hydrogen molecule were a single atom that contained two nuclei. The reason hydrogen molecules form is that the two electrons in this configuration have less energy than they do if they are in two separate hydrogen atoms. That is what chemistry is all about – minimising the energy of electrons.[1] Of course, the energy has to go somewhere, and when two hydrogen atoms collide and stick, with the electrons rearranging themselves in this way, some energy is released as electromagnetic waves (or, if you prefer, as photons), and some goes into the energy of motion of the molecules (their kinetic energy), making them move faster (which means increasing their temperature).

It doesn't always follow that because a lower energy state exists it is always going to be occupied. If I stand on top of a mountain, admiring the view, I am in a higher energy state than I would be if I were in the valley at the bottom of the mountain, because I am further away from the centre of the Earth (further out in its

[1] It is possible to make chemical reactions go in the opposite direction so that electrons end up in higher energy states, but only if there is an input of energy from somewhere. This is especially important in the chemistry of life on Earth, of course, where the source of energy is, ultimately, the Sun.

gravitational field). Provided I am careful not to fall, though, I can stay in the high energy state for as long as I like. In the case of hydrogen atoms combining to form molecules, it is very easy to get to the lower energy state (by 'rolling down the hill'), and almost any collision between two hydrogen atoms at room temperature will rearrange the electrons to form hydrogen molecules.

In a similar way oxygen atoms very easily combine with one another to form di-atomic molecules, O_2. In this case, there is one subtle difference. The outermost occupied shell of an oxygen atom contains six electrons, so it needs two more to fill the shell. As a result, using the Bohr picture, in each oxygen molecule *two* pairs of electrons (four electrons in all, two from each atom) are shared between the atoms, forming what is known as a double bond (a single bond involves one electron from each atom). Even in the quantum wave model, it turns out that the full inner shells (however many there are) play little part in bonding, and on the wave picture you can think of each oxygen molecule as consisting of two central cores (in this case, nuclei wrapped in a full inner shell of just two electrons), all surrounded, in the fat hourglass configuration, by a shell of twelve electrons.

If you started from scratch with a mixture of hydrogen atoms and oxygen atoms, and wanted to put together molecules by hand (using a *very* fine pair of tweezers!) to give the lowest energy for the electrons involved, you would make molecules of water, H_2O. Remember that each oxygen atom needs two extra electrons to fill its outer shell, while each hydrogen atom needs one extra electron. So if an oxygen atom forms bonds with two separate hydrogen atoms, the electrons in all three atoms achieve a lower energy state – lower still, it turns out, than in either oxygen molecules or hydrogen molecules. On the wave picture the two hydrogen nuclei are arranged with the oxygen core in a V-shape, with the oxygen at the apex of the V, and with a total of eight electrons (six from the outermost oxygen shell, plus one from each of the two hydrogen atoms) filling a shell surrounding all three central components.

But both hydrogen molecules and oxygen molecules are relatively stable, and at room temperature they will bounce off one another without combining to form water molecules. One way to picture why this should be so, even though the water configuration has less energy, is to think of the various molecules as occupying little hollows down the side of a hill. The hollow occupied by a hydrogen molecule and the hollow occupied by an oxygen molecule are each higher up the hillside than the hollow occupied by a water molecule. If the hydrogen and oxygen molecules were kicked out of their hollows, they would roll down the hill and into the water hollow. But they do need that initial kick to lift them out of the hollows in which they nestle – the hydrogen and oxygen molecules have to smash together fast enough for their structure to come apart, and then it will be rearranged to form water molecules.

This happens – spectacularly – if a mixture of hydrogen and oxygen at room temperature is ignited by a spark, or by a flame. The heat from the spark or flame makes a few molecules move fast enough for this rearrangement to occur. As it does so it releases more energy which heats the molecules nearby, and this triggers them to react in the same way. A blast of chemical activity sweeps through the hydrogen/oxygen mixture as an explosion, and you are left with a few droplets of water.

Because each hydrogen molecule has two atoms (H_2) and each oxygen molecule has two atoms (O_2) while each water molecule has one oxygen atom and two hydrogen atoms (H_2O), the explosion is most effective if you start with twice as much hydrogen as there is oxygen, so that two hydrogen molecules can combine with each oxygen molecule to make two water molecules ($2H_2O$), with none left over. Then the explosion is spectacular indeed – please do not try this at home!

One of the great triumphs of quantum physics was the development of a complete understanding of the chemical bond in terms of electron waves, largely by Linus Pauling at the end of the 1920s and beginning of the 1930s. The great thing about the quantum-mechanical approach to chemistry is that it is possible (at least, for the simplest molecules) to *calculate*, from the basic laws of physics, the change in energy that occurs when molecules are rearranged in the way I have just described, and the calculations predict just the changes in energy that are measured when the corresponding chemical reactions actually take place.

Erwin Schrödinger published his wave description of the electron in 1926, and within a year two German physicists, Walter Heitler and Fritz London, had used this wave equation to calculate the change in overall energy that occurs in one of the processes described above, when two hydrogen atoms get together to make a hydrogen molecule. Their results were very close to the figure derived by experiment. Later calculations, including the refinements made by Pauling, give even closer agreement between the theoretical predictions and the experimental data. This was a dramatic breakthrough in 1927. Before that, all chemists could do was say that, for reasons unknown, electrons like to pair up with one another in atoms, and atoms like to have filled electron shells. Now the equations were telling them why this should be so – that it is because those configurations have less energy, in a way which can be calculated and quantified, and matched up with the results of experiments.

In principle, the same sort of calculation can be applied for any molecules, although in practice it becomes very difficult to carry through the calculation precisely for complex molecules, and chemists have to fall back on various approximate techniques, which we need not worry about here. What matters is that the principles which underlie the behaviour of molecules are very well understood. Rather than going into details of more and more chemical reactions, which are really only variations on the themes already discussed, I want to move in the direction of the most interesting kind of chemistry – the kind that applies to living things, including ourselves. First, to set the scene, we need to know a little bit about some other forms of attraction between atoms and molecules, weaker kinds of bonding.

In some compounds whole groups of atoms work together, behaving like a single atom in a simple compound like sodium chloride (common salt, NaCl). A classic example is calcium carbonate, common chalk ($CaCO_3$). The whole CO_3 group acts like an atom carrying two extra units of negative charge (two extra electrons), and is called a carbonate ion. The calcium atom plays the same part in the calcium carbonate molecule as sodium does in common salt (NaCl), except that it carries two units of positive charge, not one (from which you would infer, correctly, that a calcium atom has two electrons in its outermost occupied shell and is happy to get rid of them if it can find a willing recipient).

The calcium atom has given up two of its electrons to the carbonate ion, which is a carbon atom joined by covalent bonds to three oxygen atoms *and* with an extra two electrons from the calcium atom. The quantum calculations, which we won't go into here, confirm that this is a relatively low energy configuration. Indeed, it is so stable that it is relatively easy to break the carbonate ion free from its partner, so that it takes part in chemical reactions as a unit. Another kind of ion which behaves in a similar sort of way is ammonium (NH_4), in which a single nitrogen atom is covalently bonded to four hydrogen atoms, but has lost one of its electrons, leaving it with an overall charge of plus one unit. Again, the reason why such a unit should be stable, in spite of its excess charge, is that it is a particularly low energy configuration.

The weakest form of attraction between atoms and molecules is the one that makes Feynman's 'little particles' sticky, 'attracting each other when they are a little distance apart'. These are the forces caused by the imperfect shielding of the nucleus of an atom (or the nuclei in a molecule) by its electron cloud. They are called van der Waals forces, after the Dutch physicist Johannes van der Waals, who investigated them at the end of the nineteenth century. They arise because in any atom or molecule the negatively charged electron cloud is spread around the positively charged core, so that a neighbouring electron cloud (surrounding another molecule or an atom) can 'see' the positive charge and be attracted by it to some extent, although the two atoms/molecules are repelled from one another once they come close enough for their two electron clouds to interact with one another.

We won't have much more to say about van der Waals forces themselves; but there is one more very important kind of bond we have not yet mentioned. This is particularly important in the story of the molecules of life; and it can be regarded as a kind of super-strong van der Waals effect, although it is weaker than the regular kind of covalent or ionic bond. It is called the hydrogen bond – because it only works for compounds containing hydrogen.

Hydrogen is unique because it is the only atom that takes part in chemical reactions which does not have at least one filled shell of electrons below the shell involved in those reactions. It only has one shell that is occupied at all, the innermost one. Helium, of course, also has only one occupied shell. But in helium that shell is full, so helium simply does not react with anything – it does not combine with other atoms to make molecules. It is quite happy as it is. Even lithium, the most hydrogen-like of the other atoms, although it does have a lone electron in its outermost occupied shell also has two electrons in the closed inner shell, shielding its nucleus to some extent from the outside world. Without its lone electron, though, a hydrogen nucleus would be naked. It would have nothing to shield its modesty, even partially; it would be unable to conceal the strength of its positive charge.

The power of the hydrogen bond that results, and its role in chemistry, can best be seen by looking at the example of water. Water, which is essential to life as we know it – the very name for a region devoid of water, a desert, is synonymous with a region devoid of life – has some very peculiar properties, and all thanks to the hydrogen bond. The most important factor in determining whether or not a substance is a solid, liquid or gas at room temperature is usually its weight – its molecular weight – found by adding up the atomic weights of all its components. The heavier a molecule is, the more energy it needs (the higher the temperature has to be) in order to move around freely enough to be a liquid, or (even more freely) a gas.

Each water molecule is made up of two hydrogen atoms (each with atomic weight 1) and one oxygen atom (atomic weight 16), with a combined weight (on the scale where carbon-12 has an atomic weight of 12 units) of 18 units – 18 daltons, named in honour of the pioneering chemist John Dalton. But water is a liquid at room temperature, even though many compounds with much greater molecular weights are gases under identical conditions – carbon dioxide, for example, with a molecular weight of 44 daltons, nitrogen dioxide, with a weight of 46 daltons, and even molecular oxygen itself, with a weight of 32 daltons. Taking one oxygen atom off a molecule of oxygen and replacing it by two hydrogen atoms reduces its weight by about 50 per cent, and yet, flying in the face of common sense, at the same time this makes the resulting water molecules stick together to form a liquid instead of flying about freely as a gas.

The way the hydrogen bond does this can be understood in terms of the geometry of a water molecule and the quantum wave description of electrons. The V-shape of the water molecule is actually quite a wide-angle V, with a separation of 104.5° (more than a right angle) between the positions of the two hydrogen nuclei in the molecule. On the electron wave picture these two nuclei (bare protons) cling on to the inner core of the single oxygen atom (a nucleus encased in a closed inner shell of two electrons), with a total of eight more electrons forming a lumpy shell surrounding all three inner components, like a large sphere with two swellings on it. But the oxygen core has a nucleus which contains eight protons, and this is by far the biggest influence on the electrons. The electromagnetic attraction exerted by each of the lone hydrogen nuclei is feeble by comparison, and the result is that the outer electron cloud concentrates at the end of the molecule where the oxygen atom is. This wouldn't be so important if the hydrogen nuclei – the protons – had some inner electrons with which to clothe themselves, but they haven't. If they had, they wouldn't be hydrogen nuclei.

To another water molecule the appearance – and effect – of a nearby water molecule depends on which end is pointing towards you. If the oxygen end is visible, what the other molecule 'sees' (or feels) is a cloud of electrons with an overall negative charge. If the end corresponding to the hydrogen atoms is pointing your way, what you see, through a thin electron veil, is a couple of bare protons – a net positive charge.[2] So when water molecules jostle against one another there is a natural tendency for them

[2] This is a good example of quantum effects at work. In order for a single proton to be visible in this way, the electron veil must actually be thinner than a single electron. But there must be some electrical shielding or the proton would form a much stronger bond. It is just as if a single electron were smeared out and stretched to become partially transparent electrically.

to link up, with the oxygen end of one molecule linking to one of the hydrogen atoms on the next-door molecule, with a strength intermediate between that of a regular covalent bond and the usual van der Waals forces. Because the angle between the two hydrogen atoms in a water molecule is so big, there is room for each of them to latch on to the oxygen end of another separate water molecule in this fashion, without those two water molecules getting in each other's way.

In liquid water the effect is transient, as bonds are constantly trying to form and being broken apart by the motion of the molecules; but it is enough to make the molecules sticky, constantly brushing past one another in a semi-slippery fashion, but unable (at least, at room temperature, and all the way up to 100°C) to break free entirely and fly off independently of one another in the form of a gas. When water freezes into ice, however, the hydrogen bonds have another profound effect. As the molecular motions slow, until they are able to form a crystal structure, with each molecule vibrating gently in its own place in the crystal lattice, it is the hydrogen bonding that determines how the lattice takes shape. The angle between the two hydrogen atoms in a water molecule is just right for the molecules to form a very open array, in which each oxygen atom is joined not only to its two regular hydrogen companions in a molecule, but by hydrogen bonds to two other water molecules; and each hydrogen atom is joined in this way to one other oxygen molecule, as well as to its regular molecular partner.

The structure that results is similar to the crystalline structure of diamond, though not as strong. It is a very open structure, with plenty of space between the atoms, and the regular pattern of the lattice array is responsible, for example, for the beautiful, regular geometry of the pattern of a snowflake. But the structure is so open that any particular amount of frozen water (ice) actually occupies a slightly larger volume than the same amount of liquid water. So ice is less dense than water, and floats on water.

This is such an ordinary, everyday phenomenon that we take it for granted. But stop and think about it for a moment. It is really utterly bizarre, as if a lump of iron dropped in a vat of molten iron were to float, rather than sink to the bottom of the vat. Most solids are denser than their corresponding liquids, because in solids the thermal motion of the atoms and molecules is slowed, and they can snuggle closer together. But in water, it is this very slowing down of molecular motions that allows the relatively delicate hydrogen bonds to form properly, and hold the molecules in place in a very open lattice. If ice had the same relationship to water as the solid forms of other compounds have to their liquid forms, ice cubes would sink to the bottom of drinks, and ice would form at the bottom of a pond or lake in winter, not as a crust on the surface of the water. The Arctic Ocean would not be covered in ice – and that would have profound effects on climate, as we shall see in chapter eight. It is all thanks to the hydrogen bond, which can be very satisfactorily explained by quantum physics, right down to the strength of the bond and the spacing of the angles in the lattice of a crystal in a snowflake.

Hydrogen is unique, because it is the only atom that has a single electron, and it is the only atom which can act as the positively charged partner when bonds form in this way. There are, though, several other atoms, as well as oxygen, which can act as the negative partner in such a link. The most important, alongside oxygen, is nitrogen. The hydrogen atom (or nucleus) can, in effect, form a bridge between two atoms which are each in different molecules – between two oxygen atoms, or between two nitrogen atoms (or, indeed, between an oxygen atom in one molecule and a nitrogen atom in a second molecule). In this way, it can hold two otherwise distinct molecules together, albeit not as strongly as a regular covalent bond would do. Without this ability, as I explain in the next chapter, we would not be here. Quite apart from the importance of hydrogen bonding in determining the properties of water, on which our life depends, it also determines the structure of the basic molecule of life, DNA. But what is it that makes molecules of life, including DNA, different from non-living molecules, such as water and carbon dioxide?

This is the most basic division of chemistry, which was known to the alchemists but still remained a puzzle until well into the nineteenth century. Some substances – even compounds such as common salt, or water – can be heated and then cooled, and always remain essentially the same.[3] Salt can be made to glow red hot, but it stays salt; water can be evaporated to make a gas, but condenses back into water again; and so on. There is another class of substances, though, compounds which do not stay the same when they are heated. Things like sugar, or wood. If you heat sugar, even fairly gently, it chars, and it does not 'unchar' when it cools down. And we all know what happens to a piece of wood when you heat it.

The distinction between the two classes of substances was formalised in 1807, by the Swedish chemist Jöns Berzelius, who was one of the earliest supporters of John Dalton's version of the atomic theory. He realised that the first group of substances were all associated with non-living systems, while the second group of substances were derived, directly or indirectly, from living systems. So he dubbed the second group 'organic' substances and the first group 'inorganic' substances. The organic compounds defined in this way were much more complicated than the inorganic compounds, with many more atoms in each molecule; and at first it was thought that they could *only* be produced by living organisms, through the action of some mysterious life force. It was only in 1828 that the German chemist Friedrich Wöhler discovered, accidentally, that he could make urea (one of the constituents of urine) by heating a relatively simple substance, ammonium cyanate, which was then regarded as being inorganic.

In the second half of the nineteenth century, it gradually became clear that exactly the same basic rules apply to both inorganic and organic chemical processes, and that the differences between the two categories are entirely due to the complexity of most organic molecules. It also became clear that there is something else that organic molecules have in common – they all contain carbon. The definition of an organic molecule was changed to be any molecule that contains carbon, while any molecule which does not contain carbon is now called inorganic.

[3] Within limits. We now know that if you heat anything enough it will be broken down into its components; but we are talking here about modest temperatures, a few hundred degrees Celsius.

Strictly speaking, by this definition even a simple molecule such as carbon dioxide (CO_2) is organic, although some chemists might regard it as an honorary inorganic compound.

The important point is that the distinction between organic and inorganic compounds, and the way they got their names, shows that there is something special about the element carbon. Carbon atoms really are the basis of the molecules of life – but as a result of their now well-understood physical properties, not because they house a supernatural life force. The key feature of the carbon atom that makes organic chemistry so complicated is that it has exactly four electrons (half a closed shell) in its outermost occupied shell (plus a single closed inner shell of two electrons). This means that it can form four chemical bonds, the maximum possible, simultaneously. If it had fewer outer electrons, there would be fewer available for bonding; if it had more, the outer shell would be so nearly full that what would matter would be the number of 'holes' in it available to bonding electrons from other atoms. Four outer electrons gives carbon maximum bonding capacity – including the capacity to bond to other carbon atoms.

Further up the periodic table there are other atoms, notably silicon, which also have exactly four outer electrons. But they have more closed shells between this outer layer and the central nucleus, so the influence of the nucleus on the outer electrons is diluted, and the bonds they form are not so strong. Carbon is the *smallest* atom that has exactly four electrons in its outer shell, so it can make four *strong* bonds at once. And that is the key to life.

Before we go on to look at the molecules of life themselves, though, there is one more feature of the way bonds form, highlighted by the way carbon combines with other elements, that is worth pointing out. It has to do with the fact that atoms and molecules are three-dimensional objects; and, once again, it is a phenomenon that can only be understood in terms of quantum physics. It is, though, an important step towards understanding how life works, and it brings home the message that our existence depends intimately on the behaviour of the quantum world.

According to basic quantum physics, the orbitals occupied by the four electrons in the outer shell of a carbon atom ought not to be all the same. There is one spherical orbital (just like the sole inner orbital that makes up the first shell), plus three other orbitals, which each have a fat hourglass shape (with one lobe on either side of the central nucleus), and ought to be arranged at right angles to each other. With just four electrons available, one goes into each orbital, although there is room in each orbital for a second electron with the opposite spin to the first one – which is why carbon can form four bonds. On this picture, when carbon does form bonds it should have three links sticking out in directions at right angles to each other, and one which has no preferred sense of direction at all. And yet, studies of things like the structure of crystals show that this is not the case.

We won't go into details, but it is logical enough that the shape and complexity of a crystal (like the shape of a snowflake) should reflect the shape of the underlying molecules of which it is composed. These kinds of studies show that when carbon forms four bonds with other atoms (in the simplest case, with four hydrogen atoms, to make one molecule of methane, CH_4), the bonds are all identical to one another, and are arranged symmetrically so that they point towards the corners of an imaginary tetrahedron, with the carbon atom at its centre.

The explanation depends entirely on a more sophisticated understanding of the quantum behaviour of the electron, and was worked out by Linus Pauling in a key paper on the nature of the chemical bond, published in 1931. The spherical orbital involved is called an *s* orbital, and the three perpendicular orbitals are called *p* orbitals. Pauling suggested that what happens is that instead of two distinct kinds of orbital operating alongside each other, the distinction is blurred, and the single *s* orbital mixes in with the three *p* orbitals to produce a set of four identical, hybrid states, known as sp^3 orbitals. This echoes the kind of quantum duality which sees an electron as being a mixture of both wave and particle. The orbitals are not either *s* or *p*, but a mixture of them in the ratio 1:3.

Of course, Pauling did more than merely speculate along these lines. He calculated the whole thing, using the quantum rules, and predicted the strength of the resulting bonds, which matches up with the strength measured by experiment. And if you want a physical picture of why this arrangement should be preferred, it is because the sp^3 structure keeps the four electron clouds in the outer part of the carbon atom at the greatest separation from each other, which corresponds to the lowest energy state available to them.

This hybridisation of orbitals can only occur for orbitals which have the same energy as each other. It is a sharing out of electrons evenly among the equivalent energy states available, and is a fundamental feature of organic chemistry, as Pauling explained.[4] It even operates in the ammonium ion, which we mentioned above. Nitrogen is the next element in the periodic table to carbon, and has five outer electrons. So it 'ought' to form three bonds, by accepting electrons from other atoms, which it does when it makes ammonia (NH_3), for example. If the nitrogen atom were to lose one of its electrons, though, it would be like a carbon atom with an extra unit of positive charge in its nucleus. Then it would be able to form a tetrahedral arrangement of four bonds, just as it does in the ammonium ion (NH_4). This is why the ammonium ion acts as a single unit in compounds where it is bonded ionically to another atom.

The capacity for hybridisation is related to the idea of resonance, in which the structure of a molecule can be regarded as flickering between two or more different states, so that it produces a kind of average over those states. This is only possible if the different states have essentially the same energy, and are different versions of the lowest energy state available.

The simplest example is the hydrogen molecule. We said before that the bond between two hydrogen atoms in a molecule (H_2) is covalent; the image offered was of a pair or electrons being

[4] It also explains how phosphorus, in the exception to the rule, can sometimes make five bonds simultaneously, although it doesn't have such an interesting chemical life as carbon. Phosphorus has an atomic number of 15, and five electrons in its outermost occupied shell. It can make five bonds by, in effect, using four of the electrons to make covalent bonds, and giving up the fifth electron to make an ionic bond. But all the five bonds it makes are identical, a mixture of one-fifth ionic and four-fifths covalent in character, in another fine example of a purely quantum process.

shared between the two nuclei to make a single bond. Then we described an alternative picture, in which the two nuclei are surrounded by a single electron cloud. In 1928 Pauling offered a third picture (a third model), in which the bond is ionic, but involves resonance. If the first hydrogen atom gives up its electron entirely to the second, it will be left with one unit of positive charge, while the second has gained one unit of negative charge, and so the two ions will stick together (H^+H^-). But exactly the same thing can happen with the roles (and charges) reversed – if the second hydrogen atom gives up its electron to the first hydrogen atom (H^-H^+). Pauling showed that the bond can be described in terms of a rapid switching between these two possibilities – that is, as a resonance.

We'll give you just a couple more examples of resonance at work before applying this, and other ideas, in describing how life works. Remember the carbonate ion (CO_3). This has an overall charge of minus two units, because it has gained two electrons from whatever atom it has bonded with ionically. So (considering only the outermost shells) it has four electrons from the carbon atom, six from each of the oxygen atoms, and two extra electrons – a total of twenty-four electrons to share around four atoms in the most energy-efficient way. How are the electrons arranged?

Remember that the chemically favoured states – the ones with low energy – are the ones in which each atom has a closed outer shell of eight electrons (or, at least, the illusion of a closed outer shell). One way in which this could be achieved would be if two of the oxygen atoms each latched on to one of the spare electrons, giving them seven electrons, and then each made a single covalent bond with the carbon atom, giving each of them a share of eight electrons in all. That would leave the carbon atom with two unpaired electrons with which to form a double bond with the remaining oxygen atom, giving both of them, as well, the illusion of a filled outer shell of eight electrons. No wonder the carbonate ion is a stable unit.

But if that were the end of the story, the carbonate ion would not be symmetric. It would have an overall negative charge at one end, and one of its bonds with oxygen (the double bond, which is, of course, stronger than a single bond) would be shorter than the other two. There are, though, three ways in which the same kind of distribution of charge could be achieved, with each of the three oxygen molecules in turn being the one to form the double bond and have no excess negative charge. Because the three possible arrangements have exactly the same energy, resonance occurs, and the carbonate ion can be thought of as rapidly flickering from one option to another, cycling round all three variations on the theme. The overall effect should be that the two units of negative charge are distributed evenly around the ion, and that the length and strength of each bond between the oxygen atoms and the carbon atom should be the same, equivalent to one and a third normal single bonds, with the oxygen atoms spaced uniformly around the carbon atom. All of this can be calculated precisely using Pauling's version of quantum chemistry, and the results of the calculations closely match the results of experimental measurements using spectroscopy and other techniques. Resonance really is a good model. And nowhere in the whole of organic chemistry is it more important than in determining the nature of the structure known as the benzene ring.

We said that carbon atoms can form bonds with other carbon atoms. Sometimes a long chain of carbon atoms can form in this way, 'holding hands' with one another, but each with two bonds left over, sticking out from the chain to link up with other atoms or molecules. Because of the natural angle between the carbon bonds, though, it is also possible for them to form a ring of six atoms, the benzene ring. It gets its name from the compound benzene (discovered by Michael Faraday in 1825), each molecule of which is made up of six carbon atoms and six hydrogen atoms. Chemists knew this long before they knew anything much about chemical bonds, simply by measuring how much carbon and how much hydrogen combine to form a certain amount of benzene. At first sight this equal mix of carbon and hydrogen atoms in a single molecule is utterly bizarre. Each carbon atom has the capacity to link up with four other atoms (so much was already clear in the 1820s), so even if six of them were linked in a chain there would still be 'room' for fourteen hydrogen atoms to be attached – as, indeed, they are in the compound called hexane. Where had the extra bonding power gone?

The answer came to the German chemist Friedrich Kekule in a kind of vision, what he called a 'waking dream', that he experienced while riding on a horse-drawn London bus in 1865. In his daydream he saw chains of carbon atoms dancing around, and suddenly one of the chains looped round and grabbed hold of its other end to make a circle. The vision led Kekule to the idea that the carbon atoms in a benzene ring are arranged in a circle (or rather, a hexagon). Each carbon atom uses three of its bonds to keep the circle closed, forming a double bond on one side and a single bond on the other side. That leaves each of the six carbon atoms with just one bond free, sticking out from the circle, with which to hold on to a hydrogen atom.

It was a brilliant insight, but it still left an awkward puzzle – until quantum physics came to the rescue. The structure proposed by Kekule for the benzene ring had alternating double and single bonds around the sides of the hexagon. But the double bonds should be shorter than the single bonds. In addition, it is relatively easy to break one of the connections in a double bond, freeing up the spare bond to take part in other interactions. But the benzene ring shows no sign of asymmetry, and none of the bonds is easier to free up in this way than any of the others.

The explanation, of course, is resonance. It shouldn't surprise you, by now, to learn that the six bonds between the carbon atoms in a benzene ring all have the same strength, one and a half times the strength of an ordinary single bond. The actual benzene ring structure is a resonance hybrid between two possibilities, one with even-numbered bonds double (starting from any arbitrarily chosen place in the ring) and odd-numbered bonds single, and the other with the double and single bonds reversed. The result is a very stable structure which forms the basis of a very large number of more complicated molecules, by replacing the single hydrogen atoms attached at one or more of the corners of the hexagon, on the outside of the ring, with more complicated groups of atoms (such as the methyl group, CH_3).

The rings can even attach to each other, one hexagon linking up with another hexagon along one of the edges, like hexagonal tiles placed edge to edge – but this really is taking us beyond the realms of simple chemistry, and into the realm of life.

CHAPTER FIVE

MOLECULES OF LIFE

The ability of carbon atoms to form rings and long chains (sometimes, long chains which have rings incorporated into their structure) is a key feature of the molecules on which life on Earth is based. The chains can be very long indeed, but to get an insight into the kind of chemistry involved, a couple of simple examples will do.

In a carbon chain, because of the angle between the chemical bonds, the carbon atoms that form the spine of the chain make a kind of zig-zag pattern, with other atoms (or groups of atoms) sticking out from the corners of the zig-zag, where they are attached to the carbon atoms. One reasonably simple compound of this kind has six carbon atoms in the chain, each attached to two hydrogen atoms sticking out from the spine. At either end of the molecule the end carbon atom is attached to a so-called 'amino' group, made up of one nitrogen atom and two hydrogen atoms (NH_2). The compound is called diaminohexane ('di' for two, 'amino' for the NH_2, and 'hexane' because it contains six carbon atoms).

A slightly shorter molecule with the same sort of structure has just four carbon atoms in the chain, and one COOH group attached at either end. It is called adipic acid. If a molecule of adipic acid meets a molecule of diaminohexane, it is easy (because it leads to a lower energy state) for the adipic acid to release an OH group from one end, while the NH_2 group on one end of the diaminohexane releases an atom of hydrogen, H. The released atoms link up to form a molecule of water (H_2O), and the two bonds left free as a result link up across the gap where the atoms have been lost, with the oxygen atom on the end of one chain linking up with the carbon atom on the end of the other chain. This makes a single new chain, eleven atoms long, with a sub-chain of four carbon atoms and a sub-chain of six carbon atoms linked by each 'holding hands' with an atom of oxygen across the gap.

If there are a lot of adipic acid molecules and diaminohexane molecules mixed together, the process will repeat, with the appropriate new bits being added at either end of the chain, to make a very long string in which this basic eleven-atom spine, and its attached atoms, repeats time after time. The result, which may contain thousands of these basic units, is an example of a kind of long-chain molecule called a polymer. This particular polymer is both common and useful – it is usually known as nylon. It is also particularly simple, as most of the atoms sticking out from the spine are hydrogen, but these can be replaced by more complex structures such as the amino group, or benzene rings, or other chains.

Benzene rings themselves are fairly flat structures, but with a ripple in the ring caused by the angle between the bonds. Again, because of the exact angle of the chemical bonds of carbon, any atoms or groups of atoms that are attached around the ring – which can occur when some of the double bonds in the benzene structure are broken and freed up to interact with other atoms – lie above and below the plane of the ring, making a three-layered structure vaguely like a sandwich. There are also variations on the theme in which one of the carbon atoms in the ring has been replaced by another kind of atom. In some of the simplest variations of this kind most of the carbon atoms in the ring are each attached on one side to a hydrogen atom and on the other to a group known as hydroxyl (OH), with the sides on which each of these attachments sit alternating as you go round the ring. H and OH together make water, of course, and this class of compounds are commonly known as carbohydrates, meaning 'watered carbon'.

The simplest carbohydrates are called sugars. The simplest of them all, glucose, is a ring like the one we have just described, with five carbon atoms and one of oxygen. Four of the carbon atoms are each attached, on the sides, to O and OH. The fifth has a hydrogen atom on one side, but a more complex group (CH_2OH) sticking out on the other.

It is very easy for these rings to link together. If two OH groups, one from each ring, come together, one is released, linking up with the hydrogen from the other OH group to make H_2O. This leaves the oxygen atom from the second glucose molecule with a spare bond with which to link up to the site where the OH group has been released from the first glucose molecule. Two of the glucose-type rings linked in this way by an oxygen atom form a molecule of another sugar, maltose.

It is also possible to make a five-sided ring molecule, just like the structure of glucose but with one of the carbon atoms and its attached side groups removed. This is known as ribose. Another five-sided ring molecule, identical to ribose except that one of its OH groups has lost the oxygen atom, leaving a plain hydrogen atom behind, is known as deoxyribose, meaning 'ribose from which oxygen has been removed'. It is the basic unit that gives its name to the most important life molecule of them all, as far as human beings are concerned – deoxyribonucleic acid, or DNA.[1]

In terms of weight, though (and leaving aside water, which makes up more than three-quarters of the weight of living things like ourselves), the most important molecules in your body are the proteins. Many proteins are very large, complex molecules; but like all complex biomolecules, they are made up of simpler units and subunits, just as nylon and maltose are made up of simpler components. The complexity of the structure of biomolecules such as proteins, and their life-maintaining abilities, have been built up and refined over vast stretches of geological time (billions of years) by evolution. Natural selection has adapted these molecules to do specific jobs, based on the way in which they form chemical bonds. More about evolution shortly.

[1] An acid, by the way, is just a substance whose molecules give up hydrogen relatively easily in chemical reactions. A base is a substance that gives up the OH group. When an acid and a base react together they produce water (H_2O) and a type of compound usually known (at least in inorganic chemistry) as a salt.

Here, we want to concentrate on what those structures are, and how they do those jobs.

As well as carbon, which is the most important atom in biomolecules, the molecules of life are strongly over-endowed with nitrogen, which is another atom that can form a variety of interesting connections (usually three at a time) with other atoms. In the rocks of the Earth's crust, the most common elements are oxygen (47 per cent by weight), silicon (28 per cent) and aluminium (8 per cent). In your body half of the dry weight is carbon, 25 per cent is oxygen, and just under 10 per cent is nitrogen. Proteins contain an even higher proportion of nitrogen than the body as a whole – about 16 per cent by weight.

The reason is that all proteins are made up from simpler units known as amino acids, and all amino acids contain nitrogen. All amino acids have the same basic structure, with four groups of atoms attached to the four bonds of a carbon atom. One is the amine group (NH_2), which gives amino acids their name. One is a carboxylic acid group (COOH), which makes the molecules acidic. A third is always a single hydrogen atom. The fourth carbon bond can attach to a variety of different chemical groups, giving different amino acids their individuality.

Chemists knew that proteins are made up of amino acids long before they knew much about the detailed structure of complex molecules, because when proteins are boiled up in a strong solution of a simple (inorganic) acid or base[2] the chemical bonds linking the amino acids to one another in a chain are broken, and what is left is a soup of amino acids, which can be investigated by conventional (if sometimes tedious) chemical means.

The common amino acids have been named after the substances in which they were first discovered, or given names which reflect some obvious property of the amino acid itself. The first amino acid to be identified was extracted from asparagus in 1806, and given the name asparagine. Another amino acid, extracted from gelatin in the 1820s, turned out to have a sweet taste, and was given the name glycine, from the Greek word for 'sweet'

[2] So the chemist is putting energy into the system, and moving the atoms and molecules out of the lowest energy state to nestle in a hollow, or series of hollows, higher up the energy hill.

(this happens to be the simplest amino acid, with a single hydrogen atom attached to the 'spare' carbon bond).

An enormous variety of amino acids could exist in theory, and very many of them have actually been manufactured by chemists. But as far as life is concerned, only twenty-three are important. Just twenty amino acids are found in all proteins, and two others are found in a few proteins. The twenty-third is really a different version of one of the first twenty. In a molecule of cysteine the basic amino acid core is attached by its free bond to another carbon atom which is in turn attached to two more hydrogen atoms and to a single sulphur atom which sticks out of the end of the molecule and has yet another hydrogen atom at its other end. The dangling hydrogen atom from one cysteine molecule can very easily combine with the dangling hydrogen atom from the end of another cysteine molecule, allowing hydrogen molecules (H_2) to escape, with the two sulphur atoms latching on to each other, forming what is known as a disulphide bond. The result is a molecule known as cystine.

The same kind of linking can occur between other amino acids, including amino acids of different types. In the case of cysteine/cystine, the two identifying groups of the amino acid join together; but in other cases the amine group from one amino acid may link up with the OH part of the COOH on another amino acid, releasing water, and with the nitrogen atom now forming a bridge (known as a peptide bond) between the two amino acid residuals. New links can be attached in the same way on either side of the resulting molecule, to make a chain known as a polypeptide. This is another zig-zag structure, in which the spine is made up of a repeating pattern of two carbon atoms, followed by one nitrogen atom, two more carbon atoms, one more nitrogen atom, and so on. And a great variety of bits and pieces (including ring structures) may be attached along the chain, depending on which amino acids have joined together to make the polypeptide.

One of the distinguishing features of this kind of chain is that the peptide bond, in which a nitrogen atom is linked to a hydrogen atom and to a carbon atom which is always attached to an oxygen atom by a single bond (it doesn't much matter what the third nitrogen bond links up with), is a rigid structure, held in place by quantum mechanical resonance. The whole chain can pivot around the other bonds, but the peptide bond cannot be twisted. As a result, the polypeptide chains can only coil up to make compact structures (like balls of string) in certain ways.[3] It was by applying the principles of quantum mechanics to such structures that Linus Pauling was able to find out how proteins coil up and to analyse their structure, opening the way for other researchers to study biomolecules in this way.

One of the most important properties of such a chain can be seen by imagining it pulled out straight, and by ignoring the order of the boring repeating pattern (CCNCCNCCN . . .) of atoms along the spine. The distinguishing feature of a particular polypeptide chain (a particular protein) is then seen as the order of the various subunits along the chain, the groups attached on the side of the spine. These attachments, sometimes known as radicals, are the bits of the different amino acids that give the different amino acids their individuality. It is the order of these radicals along the chain that, in turn, gives proteins their individuality, and makes them curl up in specific ways, allowing them (or forcing them) to take part in specific chemical reactions, but preventing them from interacting in other ways.

The variety of proteins you can make from twenty or so amino acids is enormous. The natural comparison to make is with the alphabet of the English language. There, with just twenty-six characters you can spell out an enormous number of words, including all the words in this book. If each amino acid is equivalent to a letter in a word, the number of proteins you can make by 'spelling out' the links in the chain using the amino acid alphabet is bigger than the number of words in even the largest English dictionary, because the polypeptide chains can be much longer than the average English word (haemoglobin, an average-sized protein molecule, has a molecular weight of about 67,000).

[3] In the 1980s a toy known as a Rubik snake was briefly popular; anyone old enough to remember the Rubik snake can picture the limited coiling capacity of a polypeptide chain as similar to the limited coiling capability of a Rubik snake.

But only a minority of those options are actually expressed in the proteins that are so important to life as we know it.

There are two strands to the story of the investigation of proteins. One line of attack involved determining the physical structure of the molecules (how the polypeptide chains are folded up); the other involved identifying the amino acid subunits and their order along the chain for specific proteins.

The first person to use X-rays to study the structure of crystals (at first, very simple structures, like crystals of common salt) was Lawrence Bragg, in 1912. Treating X-rays as electromagnetic waves, the way the waves bounce off the atoms in a crystal makes the waves interfere with one another like ripples on a pond, and the pattern of the interference reveals details of the structure of the crystal. Bragg invented the subject of X-ray crystallography, and received the Nobel Prize in Physics with his father William Bragg in 1915, for their joint work in this field. In the 1920s Lawrence Bragg worked out a set of rules for interpreting the X-ray patterns produced by more complicated crystals – but Linus Pauling, working on the other side of the Atlantic, worked out the same rules, and published them first in 1929. This was the beginning of a long and not always friendly rivalry between Bragg's team and Pauling's team.

The next step was to apply the X-ray technique to probe the structure of the molecules of life. It was natural to start with proteins, which, as we have seen, are more common than any other life molecules. Proteins come in two basic varieties – long, thin structures, which retain the sort of elongated structure usually associated with a chain (hair is a good example), and globular structures, in which the basic protein chain is screwed up into a ball.

The first pictures of the X-ray diffraction pattern of a fibrous protein, one of the group known as the keratins, were obtained by William Astbury (a former student of William Bragg) at the University of Leeds in the early 1930s. Keratins are found in wool, hair, and in your fingernails. Astbury discovered that there is a regular, repeating pattern in the X-ray images, which means that there is a regular, repeating structure in keratin – or rather, two different repeating patterns, one corresponding to unstretched fibres (he called these alpha-keratin) and the other to stretched fibres (beta-keratin). Although the technique was not yet good enough to reveal the exact structure of the molecules in keratin, it restricted the options available, by ruling out many possibilities. This stimulated several people (notably the teams led by Lawrence Bragg in Cambridge, and by Linus Pauling at Caltech) to try to find a way of curling up a protein chain to match the X-ray pictures.

This took a long time, partly because the researchers had to go back to basics, looking at the structure of the bonds between the individual amino acids and working out how they could (or could not) swivel around, and partly because of the Second World War, during which this research essentially came to a halt. After the war the X-ray technique became much more accurate, and this enabled the modellers to make the final push needed to determine the structure of this particular kind of protein. It was Pauling who won the race, working out the structure of alpha-keratin and publishing a series of papers in 1951 explaining how protein molecules fit together to make structures as seemingly diverse as hair, feathers, muscles, silk, and horn. The structure his team had discovered was dubbed the alpha-helix, and a key feature of the basic structure is the way the peptide bond is held rigid. Another reason for the stability of the alpha-helix is that in this particular arrangement of the polypeptide chain the NH part of one peptide bond, thanks to its rigidity, lies in exactly the right place for the hydrogen atom to nestle alongside the oxygen atom in the peptide bond four carbon atoms further down the chain, so that it can form a hydrogen bond with the oxygen atom. Every single peptide bond in the alpha-helix is tied to a neighbour in this way, and this explains why the molecule has its characteristic repeating structure, which shows up in its characteristic X-ray diffraction pictures.

The different kinds of keratin are a result of subtle differences in the arrangement of the specific amino acids included in the chains. For example, in the kinds of hard keratin that make up substances such as your fingernails, there are lots of cysteine components. But when two cysteine molecules touch, as we have seen, they release hydrogen and link together through a disulphide bridge, a true covalent bond. In hard proteins rows of alpha-helices lie side by side, strongly linked to each other in this way to form tough layers of material.

In hair the disulphide bonds operate in a slightly different way, holding sets of three alpha-helices twisted together, in the same way that strands of string can be twisted together to make a stronger rope. When the hair is treated with a chemical that breaks the disulphide bridges, it becomes soft and can be curled easily into a different pattern. Then, it can be treated chemically to restore the disulphide bridges, and will set in its new pattern – this is how hairdressers achieve a so-called 'permanent wave'.

In beta-keratins the polypeptide chains do not form helices, but lie in a zig-zag, alongside one another. Instead of hydrogen bonds forming within a chain, to hold the helix in shape, they form in a similar way between neighbouring strands of protein, making a much softer structure. Indeed, one of the best-known examples of this structure is famous for its softness – silk.

Following Pauling's success with the alpha-helix, the idea of searching for helical structures in biological molecules fired the imagination of other researchers. The biggest prize of all was to determine the structure of DNA, which was known by then (the early 1950s) to be the molecule which carries the genetic message from one generation to the next. DNA is found in the nucleus of the cells of a living organism, so it is called a nuclein. It is mildly acidic, so it is a nucleic acid, and it contains deoxyribose, so it is deoxyribose nucleic acid – DNA.

The central role of DNA in life processes began to be studied in detail in the late 1920s, with investigations of the way the bacteria that cause pneumonia do their work. After a great deal of hard, painstaking work, by 1944 it had become clear that the differences between different kinds of pneumonia bacteria were caused by differences in the DNA in their cells. In other words, it was the DNA that made different bacteria different. Since the cells of all living things contain DNA, and it had long since been established that whatever it is that makes species different from one another is carried in their cells – actually packed away in the nuclei of the cells of most species, including ourselves – it was clear that DNA held the secret of life itself. But what, exactly, was DNA? How did the molecules coil up inside the cell nucleus? How did DNA pass on information from one generation to the next?

The big breakthrough came from two researchers at Bragg's laboratory in Cambridge, Francis Crick and James Watson. At last the Cambridge group had beaten Pauling – a source of great satisfaction at the time (at least, in Cambridge), although it seems just a minor footnote of history now. And they did it using the same approach that both Bragg and Pauling had tried for proteins – a combination of studying X-ray diffraction photographs to get an idea of the kind of structures involved, and making models to work out how the various components of the molecule being studied might fit together.

By the 1930s organic chemists had been able to work out what the components of DNA were, even though at that time they still weren't fully aware of its role in life processes – for the first three decades of the twentieth century, it was generally assumed that within the nucleus DNA acted as a kind of scaffolding for some protein molecules, which were seen as the fundamental molecules of life. There are just three components to a DNA molecule, but there are many examples of each kind of component in a single molecule of DNA. The first type of component is the de-oxygenated ribose sugar that gives DNA its name – a five-sided molecule containing four carbon atoms and one oxygen atom in a ring. The second is a group of atoms known as a phosphate group, consisting of a phosphorous atom surrounded by four oxygen atoms.[4] And the third is a kind of component called a base, although there are four different bases (Adenine, Cytosine, Guanine and Thymine, usually referred to simply by their initial letters) present in molecules of DNA.

By the middle of the 1930s it was clear that DNA could be broken down into units which each contained one sugar molecule, one phosphate group, and one base. One such subunit is

[4] Incidentally, even in a DNA molecule each phosphate group carries an excess negative charge, resulting from the peculiar way the bonds are distributed (*see* note 4 of chapter four), so all DNA molecules actually carry negative charge and should, strictly speaking, be called 'molecular ions'. We won't worry about such niceties of nomenclature, since hardly anyone else does.

known as a nucleotide, and it seemed logical to guess that these subunits were linked to one another to make a chain, in a similar way to the way amino acid subunits are linked together to make a chain in protein molecules. We are talking very long chains here – we now know that there are millions of atoms in a single molecule of DNA (but note that all those millions of atoms come in just five varieties – carbon, nitrogen, oxygen, hydrogen and phosphorous – arranged in interesting patterns). But how were the nucleotides arranged to make a molecule of DNA?

A group of biochemists in Cambridge, headed by Alexander Todd, showed at the end of the 1940s that the nucleotides are, indeed, linked in a chain in a specific way. The backbone of the chain is made up of alternating sugar and phosphate groups, linked to each other, with one base group sticking out from the side of each sugar. It was this basic information that Crick and Watson used, together with X-ray diffraction pictures, to work out the overall structure of DNA in 1953.

The first X-ray diffraction photographs of DNA had been obtained in 1938 by William Astbury. But there was a gap (again, partly caused by the Second World War) before the investigation of DNA structure in this way was taken up again at the beginning of the 1950s by a team at King's College, in London. The key photographs that provided the information Crick and Watson needed were taken by a young researcher, Rosalind Franklin. Franklin would probably have solved the structure problem herself if Crick and Watson had not beaten her to it;[5] but because she died young (in 1958) she did not receive a share of the Nobel Prize that went to Crick and Watson in 1962, and therefore her part in the story is not always properly credited.

The X-ray photographs showed that the structure of the DNA molecule had to be helical. The key insight, which gained the Cambridge team a Nobel Prize, had two components. First, that the diffraction pattern meant it was a *double* helix, with two strands twisted around one another (an insight Franklin also had, and mentioned in her notebooks). Secondly, that the arrangement of bases along the spine of a DNA molecule provides

[5] Whereas Crick and Watson could not have solved the problem without her photographs.

a natural way for the two strands of the helix to be held together, just as the structure of a protein alpha-helix is held rigid by hydrogen bonds.

It all depends on the detailed structure of the molecules of the different bases. Thymine and cytosine are each made up of a hexagonal ring in which there are four carbon atoms and two nitrogen atoms, with other things (hydrogen, methyl group and so on) attached around the sides of the ring. Adenine and guanine are each made up of a similar hexagonal ring, attached along one of its sides (like two tiles stuck together edge to edge) to a pentagonal ring, like the first ring but with one less carbon atom. Again, there are a few other things, such as hydrogen and amine groups, attached around the edges. Roughly speaking, A and G are each twice as big across as C and T.

If two separate strands of DNA were laid out side by side, with the bases in the middle, and lined up so that everywhere on the first strand there was an A it happened that on the other strand there was a T, while everywhere on the first strand there was a C it happened that opposite it on the second strand there was a G, the amount of space taken up by each pair of bases (AT and CG) would be the same, so there would be no kinks in the chain.[6]

If you did line up the molecules like this, a remarkable thing would happen. The shape of the A and T molecules is just right for two hydrogen bonds to form between them. And the shape of the C and G molecules is just right for three hydrogen bonds to form between them. The two strands of DNA would be held together by the hydrogen bonds all along its length, with A fitting against T and C fitting against G, like the fit of a key in a lock, or like sets of two-pin and three-pin electrical plugs slotting into their respective sockets. The TA bridge is exactly the same size and shape as the CG bridge. The picture we have now is rather like the two parallel steel rails of a railway line, linked together and spaced apart by cross-pieces, like railway sleepers. Give the double-stranded structure a twist, and you have a double helix of DNA – *the* double helix.

[6] Each type of base (A, C, G and T) occurs on *both* strands, of course. When we say 'A opposite T' that includes the reverse situation, with T opposite A, and similarly for GC and CG.

Of course, this does not happen by chance. The structure of DNA is not made up of two random strands that get together by luck, but is built up from the nucleotide units in such a way as to ensure that A is always opposite T, and G is always opposite C. The best way to see how this happens is to look at the process by which DNA is copied when a cell replicates. It is no great trick for the chemical machinery of the cell to break the hydrogen bonds (which, remember, are relatively weak) at one end of the strand of DNA, and start to unwind it. As it does so each free end of the DNA will naturally latch on to the appropriate partners from the chemical soup of material surrounding it in the cell. Wherever a single strand of DNA has an exposed A it will latch on to a T nucleotide from the surroundings; every exposed T will grab hold of a passing A. Wherever there is an exposed C it will hook on to a G (and vice versa). As the original double helix unravels, each of the two strands builds up a new partner strand, step by step along the chain, acting as a template.[7] By the time the unravelling is complete, each strand has already completed its construction of a new partner – you get two identical DNA molecules where you used to have one. As the cell divides, one copy of each DNA molecule is passed on to each of the two daughter cells, and life continues.

It was also immediately obvious, following the work of Crick and Watson, that the sequence of bases along a strand of DNA (the pattern of the 'letters' A, C, G, and T) could carry information, like the letters of the alphabet, or the sequence of amino acids along a protein. At first sight an alphabet with just four letters in it might seem restrictive. But both the Morse code and computers use a binary alphabet, in which there are only two letters (dot and dash in Morse; on or off in the binary computer alphabet). You can spell out anything you want in a binary alphabet (let alone a four-letter alphabet), provided that you make the individual words long enough. And the DNA in the nucleus of a single cell in your body carries a lot of information –

[7] This is a slightly simplified picture. In fact, the construction of the new partner strands takes place at several sites along each original strand simultaneously, with new bits of DNA growing along the original strand and linking up with one another.

a complete description of the construction, care and maintenance of a human body.

The way DNA works is by passing on information to the cell machinery about how to make proteins. It turns out that the genetic code is actually written in words that are only three letters (three bases) long, because all that each word has to do is specify a particular amino acid. So the sequence of bases along a short section of a DNA strand might spell out a 'message' like ACG TCG TCA GGC CCT. This tells the cell machinery to arrange five particular amino acids in a certain order as it builds up a protein chain. Since a four-letter alphabet can be used to make sixty-four different three-letter words, there is no problem coding for twenty-odd amino acids in this way, and even having a few special punctuation marks, such as a three-letter word which means 'stop' (as in, 'stop forming the protein chain now'). In fact, 'stop' is such an important message that there are three ways to code it – UAA, UAG and UGA (the U stands for Uracil; its presence is explained below).

It is, of course, much easier to say 'the cell makes polypeptide chains out of amino acids in accordance with instructions coded in DNA' than it is to actually make the proteins, and the whole process is still not perfectly understood. But the key steps in the process are quite clear.

What actually happens when it is necessary to make a certain protein (and one puzzle is just how the cell knows when this is necessary) is that the relevant part of the DNA coiled in the nucleus of the cell unwinds, and the appropriate message is copied, using this as a template and building up a chain of nucleotides along the exposed strand to make another strand of nucleic acid. This new strand is not actually DNA but RNA – an almost identical long chain molecule, but one in which the sugar unit is ribose, not deoxyribose, and in which there is another base, Uracil, instead of thymine. There are still only four bases in RNA, but everywhere there would be a T in DNA there is a U in RNA (which is why U turns up in the genetic code; see above).

The RNA strand made in this way (called messenger RNA) then moves out of the nucleus of the cell into the surrounding chemical soup of the main volume of the cell, where a structure called a ribosome moves into action. The ribosome moves along the strand of messenger RNA, reading it like the magnetic head of a tape player reading the tape passing by it. It is the ribosome that interprets each three-letter word in the coded message as an amino acid, and joins the amino acids together in the right order to make the specified protein. Then the messenger RNA is broken up so that its components can be used again.

While some molecular biologists were tackling the genetic code and working out how all this works (hardly surprisingly, it took until well into the 1960s), others were working out the structures of more proteins, including the globular proteins that are made by coiling polypeptide chains. Proteins do just about everything in your body, as well as providing its structures. As far as the processes of life are concerned, the most important role is played by proteins called enzymes, all of which are globular proteins. They are molecules which encourage other molecules to interact in certain ways – in chemical terminology, they act as catalysts.

The importance of enzymes can be seen by a simple example. Imagine a large, roughly spherical molecule (the enzyme) which has two irregularly shaped indentations in its surface. One of the indentations is just the right shape to hold another, smaller biomolecule (like a piece fitting into a jigsaw puzzle); the other indentation is the right shape to hold a different biomolecule. When the two molecules occupy these holes, they are lined up alongside each other in just the right way for chemical bonds to form between them. So they pair up, and are released into the cell as a single unit, to do whatever biochemical business it is that they do.

The enzyme, crucially, is unchanged by all this, and can repeat the trick time and time again. Enzymes are like single-minded robots that repeat the same chemical task endlessly (I'm always reminded of the little brooms in *Fantasia*, endlessly carrying their little buckets of water). Some enzymes put molecules together (including the polypeptide chains) and others break molecules apart, snipping with chemical scissors at the joins between amino acids. Some carry useful molecules around to where they are needed or take waste products away to be dumped; some even carry energy from one place to another.

Enzymes are important, but enzymes are only one component of the body and, like everything else, their structure is coded in the four-letter alphabet of the genetic code, locked away in the DNA inside the cells of the body. Before we move up to the next level of size and look at how whole bodies interact with one another and their environment we want to take a brief look at the other role of DNA – how it is copied (in a slightly different way from when an ordinary cell divides in two) and passed on to the next generation.

So far we have referred to the nucleus of a cell and its larger, outer part without explaining any details of the structure of the cell. Since we have already used the word 'nucleus' in a similar way in describing the structure of the atom, the gist of our meaning is apparent without those details. In fact, Ernest Rutherford gave the nucleus of the atom its name precisely because that term had already been used in a similar context in cell biology, and he wanted to echo that terminology. In order to get to grips with reproduction, though, we ought to pick up the threads by explaining a little more clearly how the cell works.

The cell is the basic unit of life. Individual cells have all the attributes of life, including reproduction; all the complex organs of living beings, whatever their function, are composed of cells. And a fertilised egg of an animal, or the seed of a plant, is a single cell which has the capacity to divide and grow into the adult organism. This involves a lot of division and multiplication – there are about a hundred thousand billion cells in your body, roughly a thousand times more cells than the number of bright stars in our entire Milky Way Galaxy (*see* chapter ten).

Each cell is surrounded by a membrane, which encloses it and restricts the flow of chemical substances into and out of the cell. Inside this container there is a jelly-like material (cytosol), in which many different kinds of biological subunits are located and operate – for example, the chloroplasts (in plant cells) which contain the green pigment chlorophyll and are involved in photosynthesis. And at the heart of the cell, behind another membrane, is the nucleus. Everything outside the nucleus is called the cytoplasm, and it is where the cell does its work, building up the molecules of life from simple components, such as water and carbon dioxide, in accordance with instructions coded in the DNA. But we shall ignore all of that and concentrate on the DNA itself, which is stored inside the nucleus in structures known as chromosomes.[8]

When living things grow it is because cells within them are dividing in two and increasing in numbers. During this process all of the chromosomes in the nucleus are copied. Then the boundary between the nucleus and the rest of the cell disappears, with one set of chromosomes moving to each side of the cell, where they are gathered inside two new nuclear membranes. Finally, the cell divides into two daughter cells, each containing a full complement of chromosomes. It is impossible to say which set of chromosomes is the 'original' (indeed, because of the way the copying works, described above, there is no distinction between the two sets of chromosomes). Where there was one cell, now there are two, and both are as old (or as new) as each other. This process of cell division is known as mitosis.

But each chromosome carries a *lot* of DNA, and the smoothness with which the cell uncoils all this and copies it during mitosis is breathtaking. In species that reproduce sexually there are two sets of chromosomes, one of which has been inherited from each parent, which doubles the amount of copying involved. In human beings, for example, there are just twenty-three pairs of chromosomes, and between them they carry all the information about how to build a body and run it, coded in shorter sections of DNA known as genes (more about genes and evolution in the next chapter). The way all this coded information is stored in chromosomes is a masterpiece of packing.

Chromosomes are a mixture of DNA and protein – but, in a

[8] All of this describes the kind of cell which makes up the human body and the bodies of other animals and plants. Such a cell, with a well-defined nucleus, is said to be eukaryotic. Bacteria are different; they have a less well-defined cell structure, in which there is no nucleus and the DNA is not arranged in chromosomes but coiled up as a so-called nucleoid. Such cells are said to be prokaryotic, and clearly represent an earlier stage in evolution. Hardly an unsuccessful evolutionary development, though, since bacteria are still very much a part of life on Earth.

reversal of the idea people had originally, it is the protein that provides the scaffolding on which the DNA is stored. These particular proteins are members of a family known as histones, and eight histone molecules link together to make a little round ball. The DNA double helix loops twice around this ball, and is clamped in place by two other histones, which are attached on either side of the ball. Then there is another histone ball with two loops of DNA wrapped around it, and so on. Each of the bead-like balls with two loops of DNA wrapped around it is called a nucleosome, and because the short stretches of DNA linking one nucleosome to the next are flexible, the whole string of beads can be coiled up into an even more compact structure, very much in the way that a bead necklace can be coiled up into a compact space. Even these coils can be coiled up in their turn, to make supercoils.

Every cell in the human body (except the egg and sperm cells) contains forty-six tiny cylinders made up in this way. Laid end to end, all forty-six chromosomes would cover a distance of just 0.2 millimetres. But if all the DNA they store were unwound and laid end to end it would stretch over a distance of 1.8 metres, greater than the height of most people. The DNA is packed into a length only about one ten-thousandth of its stretched-out length. And yet, in amongst all this tightly folded DNA, the machinery of the cell is able to find the bit of DNA it needs when it wants to make a particular protein, unwind the section of the tightly coiled chromosome that is required, copy the DNA message on to messenger RNA, and wind everything back up neatly again. In mitosis the whole lot is unwrapped, copied, and repacked into two sets of chromosomes in the space of a few minutes. When the specialised cells involved in sexual reproduction divide, however, they carry out an even more impressive trick, known as recombination.

When sperm or egg cells are being manufactured they are produced by a different kind of cell division, called meiosis. We shall describe specifically what happens in human cells, although the process is much the same in all sexually reproducing species.[9]

[9] The only real difference is the number of chromosomes involved. There are 23 pairs in

Instead of all the chromosomes simply being duplicated, in meiosis the forty-six separate chromosomes first pair up so that each of the twenty-three kinds of chromosome is alongside its opposite number (remember, one set of twenty-three chromosomes has been inherited from each parent).

After the chromosomes have been duplicated, matching bits of DNA are then cut out of each chromosome in a pair and swapped, making new chromosomes which each contain a mixture of genetic material (a mixture of genes) from both the parents of the person in which the cell lives. This is the process called recombination. The cell then divides into two daughter cells, each with a paired set of forty-six chromosomes, but then there is a second stage of division, without any DNA copying, to produce a total of four cells which each contain only one set of twenty-three chromosomes – the 'new' chromosomes produced by recombination.

In males three of these four cells usually become sperm; in females only one develops into an egg, and the rest are discarded. The important point is that in both sperm and egg there is just one set of chromosomes, and that each of those chromosomes contains genes from each of the parents of the person in whose body the cell was made. When the sperm from a man and egg from a woman get together to make a new cell with a full complement of twenty-three *pairs* of chromosomes, that fertilised egg has the capability to develop into a new human being. In every cell of that new human being one set of twenty-three chromosomes contains a mixture of genetic information from the paternal grandparents and the other set of twenty-three chromosomes contains a mixture of genetic information from the maternal grandparents. But in every cell of the human body (except the sex cells) there is exactly the same set of forty-six chromosomes, with each chromosome carrying exactly the same genetic information as its equivalent chromosome in every other cell in that body.

people, but the number of pairs in different species seems to bear no relation to the kind of 'body' they inhabit – 7 pairs in the garden pea, for example, with 24 pairs in each cell of a potato (one more than you have!) and no less then 100 pairs of chromosomes in each cell of a crayfish.

In round terms, there are 75,000 separate genes spread out along each of the twenty-three human chromosomes (some carry more, some carry less), so sex and recombination provide for an enormous variety of potential mixtures of genes. That is why no two people are exactly alike (except for identical twins and so on, which result when an egg cell divides completely in two after fertilisation, with each of the daughter cells developing into a new human being). But because there are two sets of chromosomes in every cell, there are two versions of every gene, which adds another layer of complexity to the story.

Suppose that there is a single gene which determines a bodily characteristic – eye colour, say. This is a bit simplistic. Usually, you would expect the characteristics of a body (the details of the phenotype, in the jargon) to be produced by the interaction of several, or many, genes (the entire gene package is called the genotype); but we'll keep it simple for our example. On the appropriate chromosome inherited from one parent, the gene for eye colour may 'tell' the body to have blue eyes. But at the equivalent place on the equivalent chromosome inherited from the other parent, the instruction may be to have brown eyes. Such different versions of the same gene are called alleles. In this particular case, people with the brown-eye allele in one set of their chromosomes will have brown eyes. Only if both chromosomes carry the allele for blue eyes will you actually have blue eyes.

Because there are different alleles for virtually every gene (often several different alleles), and they are continually being shuffled down the generations into new arrangements by sex and recombination, nature is continually testing out different combinations of genes (different genotypes), producing slightly different phenotypes. At the same time, nature is also 'inventing' new genes – or, at least, variations on old genes. All of this copying that goes on when cells divide is not perfect, and sometimes bits of the DNA message get mutated. Sometimes, during meiosis, bits of DNA get cut out and lost, or put back in the wrong place, or the wrong way round. This is usually bad news. A cell with damaged DNA is unlikely to work properly, and will probably die long before it gets to the stage of forming a new kind of body – a new phenotype. But if the change is small enough it will only have a small effect, for good or bad, and can end up getting expressed in the phenotype of a new individual.

Such a change might, for example, alter the structure of one kind of protein manufactured by the body, by changing the order of amino acids along its polypeptide chain. If that particular protein were, say, haemoglobin, which carries oxygen around in our blood, the slight difference in the structure of the protein might make it either more efficient or less efficient at doing its job. If it were more efficient, the body which 'owned' that new allele would be more effective, able to breathe more easily and more likely to survive and leave offspring – and half of those offspring would carry the new allele on one of their chromosomes. If the new version of the protein were less effective at doing its job (in this case, carrying oxygen), the body would probably be sickly, always short of breath, and unlikely to survive long enough to leave many offspring.

And that, at the molecular level, is evolution. Small changes in the DNA message (a change in one letter of the DNA code could be enough to change the amino acid going into a particular location in a protein molecule) get translated into phenotypes (bodies) which are slightly different from one another, and which are tested against one another by a single criterion – which ones leave most children (that is, most copies of their own genes) behind. Sex simply helps the process along by shuffling genes around to make new phenotypes, based on different combinations of genes. But there is no 'chicken and egg' problem here. There is no doubt that the molecules came first and, as they evolved, invented all the paraphernalia of bodies (including human bodies) in order to assist their own reproduction. Biologists have an aphorism that 'a hen is the egg's way of making more eggs'. In the same way, a human being is simply the genes' way of making more copies of themselves.

Although we have approached evolution in the opposite way to which the story is usually told (bottom up, rather than top down), this is a better way to put things in their proper perspective, shifting all phenotypes (including human beings) away from the centre of the spotlight. But that really is as far as we can go with the story of the molecules of life. In order to understand how evolution works out in the world at large, we have to take the next step upwards in scale, to look at how whole bodies interact with their environment and with each other.

CHAPTER SIX
EVOLUTION

Evolution works at the level of the genes. Obviously, there must have been evolution in the mechanisms by which the cell operates, and genes, as we know them, have developed as a result of evolution. But we will not go into such details here. If you want to know why one person has blue eyes and another brown, why people are different from the other apes, why all apes are different from lizards, or why peas come in different varieties, what matters is their genes. Evolution (at the level of individual animals and plants) depends on the way genes are copied and passed on to the next generation (heredity), and the fact that this copying process is (as we have seen) very nearly, but not quite, perfect. It is the 'very nearly' that ensures that offspring resemble their parents, and are members of the same species; it is the 'not quite' that allows evolution to work, sometimes leading to the establishment of new species.

The first person to understand the way heredity works was a Moravian monk, Gregor Mendel. Mendel was a contemporary of Charles Darwin, and was carrying out his studies of heredity (in particular, investigating the way characteristics are passed on from one generation to the next in pea plants) at the same time that Darwin was refining his theory of evolution by natural selection. Unfortunately, Darwin never learned of Mendel's work,[1] and Mendel only became aware of Darwin's work late in

[1] Few people did in Mendel's lifetime, because he was not in the mainstream of the scientific world, and published his key results, in 1866, in a relatively obscure journal, the *Proceedings* of the Brünn Society for the Study of Natural Science.

his career, about the time he had given up scientific research to concentrate on the administrative duties which followed his election as Abbot of his monastery in 1868. So the two great components of the modern understanding of evolution were only put together at the beginning of the twentieth century, when Mendel's work was rediscovered and several other researchers independently came to the same conclusions as him about the nature of heredity. It was only in the first decade of the twentieth century, too, that the term 'gene' was coined, but we shall use it to describe the basic unit of heredity anyway, even when discussing the nineteenth-century work of Mendel and Darwin.

It is important to emphasise that Mendel was no ordinary monk. He came from a poor family, but was highly intelligent and thirsted for an academic career in science. The only way he could get any kind of advanced education was to join the priesthood and train as a teacher. In doing so, he studied physics in particular, and he brought to his investigation of heredity the physicists' approach, taking great care to keep his breeding lines separate, and with a thorough understanding of the correct way to interpret his results statistically – something which even physicists were only just coming to grips with in the mid-nineteenth century, and which was almost unheard-of in biology in those days. There is a touch of irony in this because a few years ago some modern statisticians, re-examining Mendel's data, claimed that his results were 'too good to be true' and that he must have fiddled his data. It turned out that these statisticians did not understand the biology properly, and had not allowed for the fact that about one in ten of the peas planted by Mendel would have failed to germinate! This story delightfully highlights the fact that Mendel himself understood *both* biology *and* statistics.

It is certainly true, though, that all his biological research had to be crammed into his spare time, when his main work was as a teacher in Brünn (now Brno); it also had to be crammed in physically, to a small plot of land set aside for him in the monastery garden. And, of course, when he became Abbot, and could, in principle, have alloted himself a much bigger acreage to devote to his research, he was too busy doing other things to do any research.

The pea studies were not the only aspect of Mendel's research, but they were by far the most important set of his experiments, and they are the ones on which Mendel's modern reputation rests. He worked with a total of about 28,000 plants, out of which, in his own words, 12,835 were 'carefully examined'. This work was chiefly carried out in the second half of the 1850s, when Mendel was in his thirties. Before, other researchers had bred large numbers of plants and studied their offspring; but they had done so more or less willy-nilly, letting the plants breed naturally and then trying to make sense of the variety of crosses that resulted. Mendel treated each plant as an individual, with its own number in his notebooks, and he kept them separate, carrying out the pollination himself, brushing pollen from one plant on to the flowers of another, so that he knew, in every case, which two plants were the parents of each plant in the next generation.

Out of the wealth of detailed information he gathered from these studies, one classic example will suffice to demonstrate the way heredity works. The great beauty of pea plants for these studies (as Mendel was well aware before he started) is that they have clear-cut characteristics which can be tracked in succeeding generations. For example, some pea seeds are green, others are yellow; some are smooth, while others are rougher, with wrinkly surfaces. Among his many experiments, in one sequence of studies Mendel took plants from a variety with smooth seeds, and crossed them with plants from a variety with rough seeds. In the daughter plants produced in this way, all of the seeds were smooth, and the roughness seemed to have been bred out. But when Mendel used only plants from this daughter generation as the parents in his next series of crossings, in the next generation (the granddaughters of the original plants), 75 per cent of the seeds were smooth and 25 per cent were rough (Mendel's actual figures were 5,474 smooth seeds and 1,850 rough seeds).

In genetic terms, the explanation is simple (although, of course, it took many more experiments for the simplicity to become clear). In the original plants with smooth seeds, each plant carried two copies of a gene (strictly speaking, two alleles) that specifies smoothness in the seeds. In the original plants with rough seeds, each plant carried two copies of the allele that specifies roughness in the seeds. In each case, there is no doubt which version of the gene is going to be expressed in the phenotype.

But in the next generation, each daughter plant inherits one allele for this characteristic from each parent. Each daughter carries one copy of the allele for roughness and one copy of the allele for smoothness alongside each other in every cell. It happens that in this case the smoothness gene is dominant. As long as the plant carries that allele, it dominates over the alternative. So in every daughter plant produced in this way smoothness is expressed (the alternative allele that is present in the genome but not expressed in the phenotype unless it is present in both copies of that gene is said to be recessive).

In the granddaughter generation, however, the genes get mixed up rather differently. Each granddaughter inherits one allele for this characteristic from each parent, and each parent has two different alleles to pass on. There is a 50:50 chance of any particular granddaughter plant inheriting either allele from one parent, and the same chance of inheriting either allele from the other parent. If we denote the roughness allele by R and the smoothness allele by S, the daughters each have a genotype (as far as this particular characteristic is concerned) that can be written in shorthand as RS. Each can pass on either an R or an S; so the next generation may have one of four combinations of this allele – RR, RS, SR, or SS (there is no real distinction, of course, between RS and SR). Three of the four combinations (75 per cent) include at least one smoothness allele, which will therefore be expressed in the phenotype. Only one variation on the theme (occurring in 25 per cent of the granddaughters) has two alleles in each cell both specifying roughness. So only one quarter of the granddaughter plants will have rough seeds.

And that really is all there is to heredity – except that thousands of different genes, many of them coming in several different varieties of alleles, are actually being passed on from one generation to the next in complex individuals such as ourselves, and that very often it is the combined effect of several different genes that determines a characteristic in the phenotype, such as your height. The absolutely crucial point is that sexual reproduction does not involve blending together the characteristics of each of the parents, but passing on genetic information in discrete units, almost like quanta. There is sometimes an appearance of blending (as when the child of a tall man and a short woman grows up with intermediate height), but this is a result of many individual genes at work, rather like the way in which a pointillist painting looks like a smoothly continuous covering of paint at a distance from the canvas, but turns out to be made up of many tiny dots when it is inspected close up.

Just occasionally, though, a gene will be copied imperfectly, and a new allele will be created. If that turns out to confer some advantage on the phenotype produced as a result, the allele will spread; if it makes the phenotype less efficient, it will die out.[2] And that is how evolution by natural selection works, as Charles Darwin discovered, even though he did not know exactly how information is passed on from one generation to the next.

Like Gregor Mendel, Charles Darwin is not always given the full credit he deserves – even though, unlike Mendel, his fame was firmly established in his own lifetime. Some popular accounts of Darwin's discovery of the idea of evolution by natural selection still paint a picture of him as a young wastrel from an affluent family who got his famous berth on board *HMS Beagle* for its voyage around the world as much by luck as on merit. But this is wide of the mark. True, Darwin did come from a wealthy and privileged background. It is also true that he neglected his official studies at university. But those official studies were first in medicine (at the behest of his father, a successful doctor), which he had to give up through squeamishness (literally running from the room to be sick the first, and only, time he started to watch an operation being performed, in those days before anaesthetics),

[2] Or, of course, the new allele may be neutral, conferring neither an advantage nor a disadvantage on the phenotype. In that case, it will hang around in the gene pool of the species indefinitely. One day, though, some change in the external environment may make that particular allele advantageous to the species, and then it will start to spread.

and then in theology, the last resort of a young gentleman in those days, grooming him for a quiet life as a country parson, but a subject of no interest to him whatsoever. Darwin's idea of neglecting his studies, however, was to take a keen interest in subjects that he was not supposed to be studying, particularly geology and botany. And it was because his tutors at Cambridge knew him to be first class in both fields (though not much good as a theologian) that he was recommended to Captain Robert FitzRoy as his companion and naturalist on the *Beagle*.

The voyage lasted from 27 December 1831 to 2 October 1836. Darwin was just twenty-two when he set out on the voyage, and twenty-seven years old when he returned to England. The round-the-world trip gave him the opportunity to observe geological forces at work, and to see how they had moulded the Earth over what must have been an enormously long period of time – far longer than most people thought in the 1830s, when the notion of a Creation event in 4004 BC, a date based on Biblical chronologies, was still widely accepted. It also showed him the variety of life on Earth in all its profusion, in many different habitats. Many other people had seen similar things before Darwin, but it was his astute mind that put the pieces of the puzzle together, and came up with an explanation of how that profusion of life had evolved. The theory, natural selection, required just the enormous span of time that was indicated by the geological record. Geology gave Darwin the gift of enough time for evolution by natural selection to do its work.

One other person did have the eyes to see the evidence before them and a mind astute enough to fit the pieces of the puzzle together. Darwin had most of his theory clear in his head, and much of it recorded in his private notebooks, before the end of the 1830s. Although he gradually leaked a few details to a close inner circle of friends, he held off from going public with the idea, not least because of his concern at how it would affect his wife, Emma. She was a Christian with firmly held traditional beliefs, while Charles was an increasingly convinced atheist. In the 1850s, though, the naturalist Alfred Wallace, who was based in the Far East, was led to the same conclusions about evolution that Darwin had reached twenty years before, and worked out the theory of natural selection in almost identical terms. Darwin was by then an eminent naturalist, with whom Wallace corresponded; so Wallace sent an outline of his theory to Darwin. It was this letter from the younger naturalist that obliged Darwin himself to come out of the closet, as it were, and to write his epic book *The Origin of Species*. It was first published in 1859, and has never been out of print since. Darwin did, as Wallace acknowledged, think of it first; but Wallace, as Darwin acknowledged, should always be remembered as the equal co-discoverer of the idea of evolution by natural selection.

That phrasing is important. By the 1850s, there was ample evidence from the fossil record that species had evolved during the long span of geological time. Although some people baulked at including human beings in the story of evolution, the idea of evolution itself was no longer either surprising or disreputable. The important contribution made by Darwin and Wallace was to provide a mechanism for evolution – the idea of natural selection. Evolution is a fact, like the fact that apples fall downwards from trees. The theory (amply supported by a wealth of evidence) is natural selection, the explanation (or model) for the fact of evolution, just as theory of gravity (amply supported by a wealth of evidence) is the explanation (or model) for the fact that apples fall downwards from trees.

Both Darwin and Wallace came at the idea in no small measure through reading the *Essay on the Principle of Population* by the Reverend Thomas Malthus, which was first published (anonymously) in 1798, and later in an extended version which carried the author's name. Malthus was impressed by the way in which populations, including human populations, have the capacity to increase geometrically, which means doubling in numbers in each timespan of a certain length. If each couple in each generation produce four children that live to maturity and have children in their turn, for example, the population will double in every generation. At the time Malthus wrote his essay this really was happening to the human population in America, where the pioneers were spreading out across the 'new' lands. The human population of North America was doubling once every twenty-five years, chiefly as a result of rapid breeding, not through immigration. But in the Old World the population stayed roughly in balance, at least in the rural communities. Why?

The same sort of question applies to all species, not just people. Even elephants, which are the slowest-breeding land mammals, could produce 19 million offspring from each original pair in just 750 years in this way. Yet the world of the 1790s was not overrun by elephants. Rather, there was just one pair of elephants, more or less, around for every pair there had been in the year 1050. A similar argument applies to oak trees, or frogs; to roses or hummingbirds; to *all* species of life on Earth. Malthus pointed out that populations are held in check by disease, by the action of predators and by limits on the amount of food available. All species, plants and animals alike, produce far more offspring than there ought to be room for. But the great majority of those offspring do not survive to reach maturity and reproduce in their turn.

Malthus saw all this in gloomy terms, arguing that the desperate conditions in which the poor of the industrialising regions of Britain lived were natural, and that starvation and disease were the natural, and therefore 'right', mechanisms for holding the population in check. If the conditions of the poor were improved, they would only breed until starvation and disease once again kept their population in check; so, it was argued, leaving things as they were so that nature took its course would actually mean fewer people starving and suffering from disease.[3]

Darwin and Wallace saw beyond this superficial argument. They realised that the surplus of young individual members of a species, compared with the number that actually get to reproduce, means that there is an intense struggle for resources, a struggle for survival, which goes on *between the individual members of the same species*. As Darwin wrote in one of his notebooks in

[3] Astonishingly, two hundred years after Malthus' essay first appeared, a similar argument is still occasionally put forward today by ignorant people as a reason for not providing aid to poor countries. But the way out of the Malthusian trap is simple today – the availability of effective contraception means that human populations no longer increase automatically when living standards rise and a greater proportion of the babies born live to maturity. People simply have fewer children, safe in the knowledge that the ones they do have will not die in infancy.

the autumn of 1838, the day he read Malthus' *Essay* for the first time:

> On an average every species must have same number killed year with year by hawks, by cold, &c. – even one species of hawk decreasing in number must affect instantaneously all the rest. The final cause of all this wedging must be to sort out proper structure ... there is a force like a hundred thousand wedges trying to force every kind of adapted structure into the gaps in the economy of nature, or rather forming gaps by thrusting out weaker ones.

Species are adapted to what Darwin called 'the economy of nature' by fitting into what are known as ecological niches. A carp is good at fitting into a watery niche, and a bear is good in a particular land-based niche. But even if a bear catches a particular member of the carp family and eats it, the bear and the carp are not in competition with one another. To each of them, the other is simply part of the environment, like the weather. A hypothetical carp that developed a sense that made it turn away from the bank and dive deep when a bear approached would do well *in competition with other carp* that lurked near the bank and got eaten. A bear that was a particularly skilful fisher of carp would do well *in competition with other bears* that were less skilful and went hungry. In each case, the genes that made that individual carp, or that individual bear, successful would spread, because in each case the individuals would be more likely to live long and prosper, leaving more offspring behind them than other members of the same species.

This is what fitness, in the Darwinian sense, is all about. Not the fitness of an athlete, involving physical strength and agility (although that may come into the story), but the fitness of a key in a lock, or a piece of a jigsaw puzzle into the overall picture. Individual species, and individual members of that species, *fit into* their ecological niches.

Throughout most of the history of life on Earth, this means that evolution works mainly to keep species adapted, ever more finely, to their niches. The raw material for evolution comes from the variability in heredity that we have already discussed, as genes are copied (sometimes imperfectly), shuffled, and passed on from one generation to the next. Because the copying of DNA is not always perfect, evolution still occurs in species, such as bacteria, which reproduce asexually – albeit more slowly in terms of generations, which is just as well, or we would not be here. But the competition between individuals ensures that there is a natural selection of individuals from the variety assembled in each generation, with only the fittest for the environment in which they live surviving best and leaving most offspring. If, for example, a longer beak helps a hummingbird to get more nectar and survive to reproduce, then in each generation the birds with longer beaks will be favoured, if only slightly, over birds with shorter beaks. Over very many generations, the average beak length of members of that species will increase.

So where do new species come from? The full power of evolution by natural selection at work is seen when the environment changes, or when individuals move into a different environment. The classic example, which Darwin saw in the Galapagos Islands and which helped to shape his thinking about evolution, is the variety of different kinds of finch found on the different islands.

The birds on the different islands are all recognisably finches, but on each island the species is superbly adapted to the life it leads there. There are many islands, and many varieties of finch; but we will give just one pair of examples. The food that is available is different from one island to the next. On one island a long, thin beak is useful to probe and pick at the food available. On another island the seeds that are available are best dealt with using a broad, heavy beak to crack them open. On each island the finches have just the kind of beaks needed to make best use of the food resources available.

It was clear to Darwin (and has since been proved by molecular studies of the DNA itself) that the Galapagos finches were all close relations, descended from a few original birds that had arrived there from the mainland. Over many generations the Malthusian pressures of the struggle for survival had produced different species from the original stock. Given even more time, Darwin argued, the same process could explain the evolution of all life on Earth (including humankind) from a common ancestor.

There are still people around who think that evolution is 'just a theory', in the same sense that your Uncle Arthur might have some crackpot idea about how to grow better roses. 'Where's the proof?' they ask. Proof there is, in plenty, but it has largely been buried away in technical journals and books, not readily accessible to doubting Thomases. But there is one book in particular, Jonathan Weiner's *The Beak of the Finch*, that describes evolution going on, in exactly the way that Darwin surmised, literally before the eyes of biologists. Deliciously, the species that can be seen evolving as they adapt to changing environmental conditions are the finches of the Galapagos islands, first made famous by Darwin himself.

The story of the twenty-year research programme that revealed evolution at work is so spectacular that it hardly needs dressing up. Rosemary Grant, her husband, Peter, and their colleagues have returned season after season to the Galapagos islands since 1970, and literally know every one of the finches on one of the islands by sight. They have kept family trees of the birds going back all that time, and know which birds have bred successfully and which have not. And they have trapped all but one or two of the birds, measuring and photographing them before releasing them back into the wild. They have watched populations decline in times of drought and boom in times of plenty; and they have seen how for one particular species a change in the length of beak of a bird of less than a millimetre can make all the difference between it thriving and leaving many descendants or dying before it has a chance to reproduce. Only the biggest birds with the biggest beaks, able to cope with the toughest seeds, made it through the worst droughts.

The story is brought bang up to date, in technological terms, with the work of Peter Boag, who has studied the DNA in samples of blood from the finches, and can actually see the differences in the genetic code which specify the different designs of beak *and* make one finch good at eating seeds while another specialises in sipping nectar from cactus flowers.

Weiner also discusses the continuing resistance to Darwinian ideas, and the extent to which even among scientists non-biologists sometimes fail to comprehend how evolution works.

Chemists who invented new pesticides were astonished when populations of insects developed resistance to them, but the whole point about evolution is that any new method of killing things will, unless it completely wipes out a species, give rise to a population resistant to the killer. One result is that cotton farmers in the very States of the US where resistance to Darwinian ideas is at its strongest are struggling every season with the consequences of evolution at work in their own fields.

In hospitals, bacteria that cause diseases are increasingly resistant to drugs such as penicillin, for the same reason. The drugs kill all the susceptible bacteria, but, by definition, the survivors of an attack are the ones who are not susceptible to the drug. The more you kill susceptible bacteria, the more opportunities you give the others to spread – and bacteria breed far more quickly than human beings, offsetting the disadvantage they would otherwise have by not breeding sexually. The surprise, to an evolutionary biologist, is not that after half a century of use penicillin is losing its effectiveness, but rather that it retains any effectiveness at all.

But *why* are some people so hostile to the idea of evolution? Perhaps it isn't the idea they are hostile to at all. In one of the most telling passages of his book, Weiner describes how one of the scientists involved in the research told of a long plane flight during which he got chatting to his neighbour, and described in detail what his work was all about.

'The whole time on that plane, my fellow passenger was getting more and more excited. "What a neat idea! What a neat idea!" Finally, as the plane was landing, I told him that the idea is called evolution. He turned purple.'

The Beak of the Finch is not the best place to learn about evolution if you have an open mind – Richard Dawkins' books fit that brief. But it is the ideal book to recommend to any doubter who asks, 'Where's the evidence for evolution?', and it is an entertaining insight into one of the most important pieces of biological research of the past twenty years, which is why I have elaborated on Weiner's themes at such length here. It literally describes evolution in action, operating in just the way that Darwin and Wallace suggested.

In the Galapagos, different species have evolved because their ancestors moved to a new environment. But sometimes, during the long geological history of the Earth, it is the new environment that has, in a sense, 'come' to the species, and forced them to evolve or die. Several times in the history of the Earth there have been events in which many species were wiped out entirely, and the survivors radiated and evolved to fit the new ecological niches available. The most famous, and the most relevant to ourselves, is the event sixty-five million years ago, almost certainly caused by the impact of a comet or asteroid with the Earth, which wiped out the dinosaurs and left the way free for mammals to radiate into new ecological niches and to evolve new forms, including (eventually) *Homo sapiens*.

Mammals had been around for more than a hundred million years before the catastrophe that brought an end to the era of the dinosaurs occurred, at the end of the Cretaceous period of geological time, some sixty-five million years ago. But because the dinosaurs were so successful, the opportunities for mammals were limited. There were dinosaur equivalents of large grazers, like modern elephants and deer, and there were dinosaur equivalents of modern predators, like lions and wolves. Mammals were restricted to the roles of small, shrew-like creatures, scurrying about in the undergrowth, largely living off insects. But the 'terminal Cretaceous event' wiped out all the large species, and left their ecological niches empty. So there was scope for the mammals to spread out, ecologically speaking, to fill the roles left vacant by the absent dinosaurs. It took just three million years for some of the shrew-like mammals of 65 million years ago to evolve into creatures the size of dogs, with bats, rodents and hoofed animals following hot on their evolutionary heels. By fifty million years ago, ancestral elephants, the size of modern pigs, had appeared.

The speed of this early radiation and adaptation of the mammals at that time was caused largely by the variety of opportunities they were offered by the death of the dinosaurs. The processes which specifically led one mammal line, the primates, to produce ourselves were also affected by changes in the geography of the planet, as the continents shifted position, and changes in the climate, which came partly as a result of those geographical changes – more about these changes affecting the whole Earth in the next two chapters.

Our own place in evolution, and in particular our relationship to our nearest cousins, the African apes, is shown not only by the fossil record, but also by direct comparisons of the DNA in our bodies with the DNA in theirs. This shows that more than ninety-eight per cent of the DNA in human beings, gorillas and chimpanzees is the same – the differences that make us uniquely human amount to only a little over one per cent of our DNA. Studies of the DNA molecules of many species have been used to work out how rapidly changes in DNA have occurred during the recent course of evolution, and these show that the three-way split leading to modern humans, chimps and gorillas must have occurred about five million years ago, at a time when the East African woodlands that our ancestors inhabited were drying out and in retreat, forcing those ancestors to adopt new lifestyles and adapt to changing conditions in order to survive.[4]

Human beings have appeared almost exactly halfway through our scientific overview of the Universe, and this is no coincidence. It results from the way we have chosen to look at things on different distance scales, starting small and working upwards. An atomic nucleus has a radius of about 10^{-15} m, and a human being has a length of about one metre. So a person is about 10^{15} times bigger than an atomic nucleus. A light year is about 10×10^{15} m. So when we look to distances as much bigger than a person as a person is compared to a nucleus, we are out in the Universe at large, in the realm of the stars. In this sense, people are roughly halfway in size between the world of nuclei and particles and the world of stars.

People are also just about as large as it is possible to be and still lead an active lifestyle on the surface of the Earth. Remember that it needs the gravitational pull of the entire Earth to break the electric bonds that hold an apple to a tree and send it falling to the ground. But gravity at the surface of the Earth is, indeed, just strong enough to break electric bonds in this way. Similarly,

[4] Happily, this molecular dating is in good agreement with the fossil evidence.

if you fall over and break a limb, the break is caused by the fact that the gravitational pull at the surface of the Earth, which gives you your weight, is strong enough to disrupt the electric forces that hold the atoms and molecules in your bones together. Children fall over often, without doing themselves much harm, because they are built lower to the ground and do not have so far to fall, but anything much more than two metres tall is in serious trouble if it falls down. The only way for a mammal to get much bigger is to be sturdy and ponderous (like an elephant) or to be supported in water (like a whale).

There is, though, another way of looking at human size in relation to the Universe. Most important of all, in terms of the theme of this book, people are the most complex systems that we shall encounter – indeed, they are the most complex systems in the known Universe. And this, too, is largely because of the trade-off between gravity and the other forces of nature. In round terms, as we have mentioned, there are a hundred thousand billion cells in your body, all working together to make one living organism. They work entirely through electromagnetism, which underpins all chemical reactions. The number of cells in the body makes both for complexity and for the possibility of specialisation, with many cells adapted to work together to do specific tasks, including the task of forming a large and complex brain. But all the interesting things that we do are a result of chemical processes, driven by electromagnetic forces. As we shall soon see, once you start looking at objects on a larger scale, such as planets and stars, gravity crushes all (or at least, a great deal) of the interesting electromagnetic structure out of existence. A planet contains more atoms than a human being does, but it does not have such a complex structure.

On the scale of nuclei and particles, things are relatively simple, because only a few particles are involved in interactions at any one time. On a human scale, things are complicated and interesting because it is possible for what is really a rather delicate structure of a hundred thousand billion cells to get together and interact as one unit, and for molecules as complex as DNA to carry out their work. But things like planets and stars are also relatively simple, because on a planetary scale and upwards molecular complexity is destroyed by gravity and we are back to the simplicity of inorganic chemistry, at best. Inside a star, even that amount of complexity is not possible, and we are right back to the simplicity of particle physics.

More of all this in its place. First, I want to describe some of the interactions which involve the most complex systems in the known Universe – ourselves. People are far too complicated to be explained by the kind of scientific rules which make the orbit of the Moon around the Earth, say, so predictable. It is not possible to say exactly how an individual human being will respond to any particular outside influence when it comes to conscious behaviour, rather than something simple like falling under the influence of gravity. Evolution by natural selection is, though, amply powerful enough to explain why we behave the way we do in general terms, addressing such questions as how and why people choose their sexual partners, why altruism should be evolutionarily successful, and why there is conflict, in many families, between parents and their teenage offspring.

Charles Darwin was the first person to apply evolutionary arguments to attempt an understanding of human behaviour, and wrote (as early as 1839, although the essay was not published at that time):

> Looking at Man, as a Naturalist would at any other Mammiferous animal, it may be concluded that he has parental, conjugal and social instincts, and perhaps others. – The history of every race of man shows this, if we judge him by his habits, as any other animal. These instincts consist of a feeling of love <& sympathy> or benevolence to the object in question. Without regarding their origin, we see in other animals they consist in such active sympathy that the individual forgets itself, & aids & defends & acts for others at its own expense.

This idea of 'looking at Man, as a Naturalist would at any other Mammiferous animal' is at the heart of the modern understanding of evolution. People are animals, and have been shaped by the same sort of evolutionary forces as other animals. The study of all forms of social behaviour in all animals, including humans, in this way is sometimes known as sociobiology, and it is one of the triumphs of evolutionary theory that the modern understanding of genetics and heredity, combined with the idea of natural selection, can explain the origin of the altruistic behaviour (in humans and in other species) that Darwin himself puzzled over in 1839.

We do not have room to give more than a caricature of how sociobiology works, but it should at least convince you that there is something in the idea.[5] So-called altruism crops up in many ways in nature, but here are two examples. First, why does a bird in a flock that is feeding on the ground give a warning cry when a predator approaches? You might think that by doing so it would draw attention to itself, and make it more likely that it will be killed and eaten, so that it fails to pass on its genes to the next generation. Secondly, why is it that human beings are sometimes motivated to risk their own lives to aid a stranger – as when somebody dives into a river to rescue a drowning child? Again, this activity does not look, at first sight, like a good way to ensure that you pass on copies of your own genes.

But in a flock of birds the chances are that all the birds are related to one another, and that many of the genes carried by one individual will also be present in many other individuals in the flock. If one of those genes (or several working together) encourages an individual to give a warning cry which saves the others, even if that individual is eaten, many copies of that gene package will survive in the bodies of the birds that fly away to breed another day.

The case of the drowning child rescued by an unrelated bystander is a little more complicated, but shows how our behaviour is still shaped by our evolutionary past, even though society has changed dramatically in recent centuries.

Until quite recently, in evolutionary terms, most people lived in tribes or villages. So, if you saw a child in danger, there would be a good chance that it was related to you.[6] A gene package which encouraged an instinctive reaction to jump to the rescue

[5] If you want to know more, see *Being Human* by Mary and John Gribbin (Phoenix, 1995).

[6] Perhaps equally significantly, in a small community, if you helped someone in this way, there was a good chance that, if you were in danger later in your life, they might be around to help you. This would reinforce the argument outlined here.

would spread amongst the population of the tribe or village, provided that the benefits (in terms of the increased chances that gene package gets to spread if the child survives) outweigh the costs (in terms of the lost chance of the gene package to spread if the rescuer drowns). This works best if the drowning child and the rescuer are closely enough related that there are, indeed, many copies of genes that they both carry, including this 'altruism' gene. So, what are the chances?

Very many genes are, of course, carried by very many individuals (like the allele for blue eyes). But if we want to get personal, since you inherit half of your genes from your father and half from your mother, there is a 50:50 chance that any one gene in your mother is also present in you, and similarly with your father. There is also a 50:50 chance that you share a particular gene with one of your siblings, and a one in eight chance that you share any particular gene with a first cousin (the offspring of your mother's sister), and so on. This led to a remark that J. B. S. Haldane is alleged to have made in a pub in the 1950s. Discussing the problem of altruism over a beer with some colleagues, he was asked if he would lay down his life for a brother. After barely a moment's thought, he replied, 'Not for one brother; but I would for two brothers, or eight cousins.'

The point is that, on average, any behaviour that ensures the survival of two of your siblings, or eight of your cousins, ensures the survival of *all* of your own genes! In real life you don't necessarily have to lay down your own life to rescue a brother or a cousin, or a stranger, even if you jump into a river to rescue them from drowning. There may be an *element* of risk, but also a good chance that you will both survive. In statistical terms, if there is a better than 50:50 chance that you will survive, it is worth the risk to yourself to save a drowning brother, and so on.

Now, nobody is suggesting that people stand around on the banks of rivers calculating how closely related they are to the drowning child and then calculating the chances of success, before jumping to the rescue. There is a range of human reactions to the crisis, just as there is to most things. Some people pass by on the other side, some dither, some jump in without thought. The point is that this whole pattern of individual responses has evolved to take account of the genetic odds that existed in the days when we lived in smaller communities. Genes that made people too reckless died out, of course; but, to some extent, genes that made people too cautious also died out.

The crucial balance that has emerged is that in most actual rescue attempts the rescuer should have a good chance of surviving, while the drowner should have very little chance of surviving unaided. Then, the balance works in favour of the spread of the gene (or genes) that encourages this kind of altruism. Altruism is, in fact, a shining example of what Richard Dawkins has called 'the selfish gene' at work. 'Selfishness', in this case, means ensuring that copies of the gene survive, and it doesn't matter which body they survive in. So genetic selfishness actually makes us, in some circumstances, *un*selfish at the level of the individual person. Without this genetic selfishness, it would seem totally unnatural to us that anybody should ever risk their own life to save a drowning child.

What about sex? Why is the ratio of males to females, in the human species and other mammals, so close to 1:1? After all, a single male could, in principle, impregnate many females and produce many offspring. The species doesn't 'need' as many males as it does females in order to maintain its numbers from one generation to the next. But look at it from the point of view of the genes. Imagine a species in which each successful male has a harem of ten females, and no other males get to reproduce at all (not so different from the actual case in species such as red deer). You might think that the ratio of males to females at birth in such a situation ought to be 1:10, so that no effort is wasted by mothers raising males who will never reproduce in their turn. Surely, the mother would be better off raising more females, who will be certain (provided they reach maturity) to pass on copies of her own genes to later generations? But every male that does breed produces ten times as many offspring as each female. So, if a population ever did exist in which each mother produced ten daughters for every son, a mutation that caused one female to produce a few extra male offspring would be at an advantage, because those sons would breed very successfully – more successfully than each of her daughters.

The mutation would spread, and natural selection would ensure that the pattern we actually see in species such as deer evolved. The number of male and female offspring born in each generation is the same, because although the males have (in our hypothetical example) only a 1 in 10 chance of reproducing, each male that does reproduce produces exactly ten times as many offspring (ten times as many copies of its genes) as each female. For obvious reasons, this is called an evolutionarily stable strategy, or ESS; any mutation which makes mothers consistently produce *either* more females *or* more males will be at a disadvantage, and will die out. There is always a 50:50 chance, at conception, that the offspring will be male, and a 50:50 chance that it will be female. Which doesn't mean that individual mothers might not have a string of sons or a string of daughters, just as the 50:50 chance of tossing heads on a coin doesn't stop the occasional string of heads (or tails) coming up if you toss the coin enough times.[7]

Similar arguments explain the battle of the generations that goes on in so many households. Since each parent passes on half of its genes to each child, on average a couple needs to produce two children that survive to maturity to ensure that all their genes are passed on. Obviously, it would be a good thing, in terms of their own selfish genes, to produce more children, as a kind of backup, in case anything happens to the first two. But this is only a good thing if the effort of raising the 'extra' young doesn't make the parents so exhausted that they neglect their firstborn, which dies in infancy. Each couple, though, only really 'cares' (in terms of evolution) about the survival of its own genes. As the firstborn gets older it takes less effort to raise it, and there is less chance that it will die if effort is diverted into raising more young. As soon as the offspring are able to fend for themselves,

[7] Sex is actually determined, in humans and similar species, by a gene carried on a particular pair of chromosomes. All females have an identical pair of these, labelled XX, so each mother must pass on X to each of her offspring. All males have a non-identical pair, labelled XY. Since they pass on one of the pair, selected at random, to each of their offspring, there is exactly a 50:50 chance that the offspring will be XX or XY. In evolutionary terms, it would be quite easy for a mutation to ensure that a male passed on all Xs or all Ys to its offspring. The fact that such a mutation has not spread in the human gene pool confirms that the 50:50 balance of the sexes is an ESS.

the best thing, in terms of the parents' evolutionary success, is to get rid of the offspring and raise another infant. From the offspring's point of view, though, the best thing is to hang around and continue to get help from the parents. It happens in birds, and it happens in people, especially in more 'primitive' societies. In many modern societies the instinctive conflict remains, even though, thanks to our modern cultural surroundings, the parents may have no intention of actually producing more offspring.

The fact that the simple evolutionary patterns are complicated by the development of modern society is one reason why the application of sociobiology to human beings is still far from easy, and sometimes controversial. We think about our actions (at least, some of the time) instead of always acting on instinct; but since thinking has evolved, and is obviously successful in evolutionary terms, that too is, in principle, a valid subject for sociobiological investigation. People are, as we have said, very complex creatures – the most complex single systems that we know about. Having brought you from the simplicity of the subatomic world to this peak of complexity, it is time to move on, to still larger scales, and to look at the whole of our planet and its place in the cosmos. The good news is that as we move out into the Universe at large, the basic science will, from now on, get easier to understand once again.

Albert Einstein once commented that 'the eternal mystery of the world is its comprehensibility . . . the fact that it is comprehensible is a miracle'. But we can now see that the reason the Universe – what Einstein meant by his use of the term 'world' – is comprehensible is because it is simple. The reason for its simplicity may still be mysterious; but because it is itself simple, creatures as simple as ourselves (complicated though we are by the standards of the Universe) are able to understand it. We will begin our journey into space by looking at the solid Earth beneath our feet – which is not always as solid as you might have thought.

CHAPTER SEVEN
OUR CHANGING PLANET

By human standards, the Earth is big. It has a diameter of nearly 12,800 km, and a circumference, in round terms, of 40,000 km. But most of the things that affect us directly take place in a very thin layer near the surface of the Earth – the crust of the solid planet and the thin blanket of atmosphere that surrounds it. To put this life zone in perspective, imagine slicing through the solid Earth, like a cook slicing an onion in half, and taking a look at the layered structure that lies beneath our feet.

We know a little bit about the internal structure of the Earth thanks to the vibrations caused by earthquakes – seismic waves which travel through the rock and are reflected and refracted by boundaries between different layers of rock in a way that is analogous to the way light is reflected and refracted from the surface of a block of glass. Just as a physicist could work out some of the properties of the glass in a prism by shining light through it and studying the way it is refracted, so geophysicists can work out some of the properties of the Earth by monitoring seismic waves. But this is a very crude kind of 'X-ray' process, which depends on naturally occurring earthquakes to make the seismic waves, which are not under any sort of human control, and on having an array of seismic stations around the globe to monitor the seismic waves when they are produced. In many ways, we know more about the stars in the sky than we do about the centre of the Earth – after all, we can see the stars!

The deepest that any person can get below the surface of the Earth is to the bottom of the deepest mine, a mere 4 km; and the deepest borehole ever drilled into the Earth's crust reached less than 20 km below the surface. Although the details of the Earth's gravitational and magnetic fields give some extra information about what is going on inside the Earth, for the most part we are still dependent on those seismic waves.

What they show is a layered structure built around a solid inner core, which has a radius of about 1,600 km (the exact radii that we quote here are all a little vague, reflecting the difficulty of making seismic observations, so don't be surprised if you see slightly different figures in different books). This inner core is surrounded by a liquid outer core, which has a thickness of just over 1,800 km. The whole core is very dense, probably rich in iron, and has a temperature of nearly 5,000 °C. The circulation of this electrically conducting material in the liquid outer core is clearly responsible for the generation of the Earth's magnetic field, but nobody has ever been able to work out a completely satisfactory model of how this process works.

The high temperature in the core is partly left over from the way the Earth formed, as a hot ball of molten material made from many smaller objects that collided and stuck together when the Solar System formed (more of this in chapter nine). Once a cool crust had formed around the molten ball of rock it acted as an insulating blanket, trapping the heat within and only allowing it to escape slowly into space. Even so, without some continuing injection of heat the interior of the Earth could not still be as hot as it is today, more than four billion years later. The extra heat comes from radioactive isotopes (originally manufactured in the death throes of stars), which decay into stable elements and give out energy as they do so. In about ten billion years time even this source of heat will be used up, and the Earth will gradually cool down and freeze into one solid lump; but by then, the Sun itself will have died, so the freezing of the interior of the Earth will be the least of the worries of any intelligent beings around to witness the event.

The bulk of the volume of the Earth is made up of the layer above the core, which is known as the mantle. The entire mantle is just under 3,000 km thick, but is usually regarded as having two components (distinguished by slightly different seismic properties), the lower mantle (about 2,300 km thick) and the upper mantle (some 630 km thick). Together they account for 82 per cent of the volume of the Earth, and two-thirds of its mass. Above the mantle, like a skin around the solid Earth, is the crust. This is about 40 km thick under the continents of our planet, but only 10 km thick underneath the oceans. Taking an average of 20 km, in round terms the thickness of the crust represents less than one-third of one per cent of the distance from the surface to the centre of the Earth.

Geophysicists know much more about the structure of the Earth near the surface than they do about the core (but 'much more' is still not a lot). Between depths of about 75 km and 250 km there is a region of the mantle in which seismic waves travel slightly more slowly than they do in the regions just above and below this zone. This is a zone of weakness, caused by a partial melting of the rock in the low-velocity zone, which is called the asthenosphere (the bulk of the mantle below the asthenosphere is called the mesophere). The rocks can melt partially in this region because of a trade-off between pressure and temperature (the same sort of trade-off explains why the outer core is molten but the inner core is solid). Although the rock is hotter deeper into the mantle, the pressure is also greater, and the rock is solid; and although the pressure is less above the asthenosphere, so is the temperature, and once again the rock is solid. In the asthenosphere, the combination of pressure with the temperature of about 1,100 °C, the temperature inside a blast furnace, is just right to make the rocks slushy.

Everything above the asthenosphere is called the lithosphere, and the crucial point is that because of the weakness of the asthenosphere, pieces of the lithosphere (which is about 100 km thick, on average) can move about, in a sense floating on the asthenosphere. It is this freedom of pieces of the lithosphere to move about on the asthenosphere that makes the continents themselves move about (very slowly) on the surface of the Earth, changing the geography of our planet dramatically over the course of geological time. But it is only relatively recently that the way the continents move has been understood – indeed, it is only relatively recently that most geologists and geophysicists accepted the evidence for continental drift at all, although a few people had long been persuaded that the continents do move about on the surface of our planet.

Speculation about the causes of the present arrangement of the continents has been rife since the earliest days of reliable mapping on a global scale. In 1620, little more than a century after Christopher Columbus made his famous voyage of discovery, Francis Bacon drew attention to the similarity in outline of the eastern seaboard of South America and the western coast of Africa when he wrote:

> Each region has similar Isthmuses and similar Capes, which is no mere accidental occurrence. So too the New and Old Worlds are conformable in this, that both Worlds are broad and extended towards the North, but narrow and pointed towards the South.

There is no evidence that Bacon actually thought that these similarities were a result of the two continents having once been a single land mass that broke apart and were shifted across the globe, but if a cutout of South America is moved around on a globe (or such a rearrangement is simulated in a computer), the bulge of Brazil does indeed nestle neatly in under the bulge of west Africa. North America, if given a bit of a twist, fits almost as well up against Europe, with Greenland bridging the gap between them in the north.

The first published version of a map showing this fit came from Antonio Snider, an American working in Paris, in 1858. His idea was that when the Earth cooled, the continents had formed as a single land mass on one side of the globe, and that because this was an unstable arrangement the single supercontinent had split and the fragments moved apart, being carried rather rapidly to their present-day positions in a catastrophic process which Snider linked with the Biblical story of Noah's flood.

Variations on the theme were discussed, in a quiet way, by several people in the following decades.[1] But the person who has a rightful place in history as the 'father' of the idea of continental

[1] Including, incidentally, the astronomer George Darwin, one of the sons of Charles Darwin. A reminder that we had a good theory of evolution long before we had a good theory of continental drift.

drift is the German Alfred Wegener, who, unlike other people who speculated about the possibilities, worked out a detailed model – which was not entirely right, but that is beside the point since few models are – and kicked up a fuss about his ideas, making a positive effort to convert other geologists, instead of just publishing quietly and leaving the theory to sink or swim without help. Wegener lived from 1880 to 1930, and first published his ideas on continental drift in a monograph in German in 1915. The book remained largely unknown outside Germany, because of the First World War, and it was the third (greatly expanded) edition, which was published in 1922 and translated into English in 1924, that really started the modern debate on the subject.

What now seems, with hindsight, like key evidence that continents which are now on the opposite sides of the oceans were once joined together comes from geology. Rock formations in, for example, west Africa join up with rock formations in Brazil as clearly as if the outlines of the two continents were drawn touching each other on a sheet of newspaper, and then the paper was torn along the join and the pieces moved apart. The original fit could easily be reconstructed by placing the two pieces of paper together so that the lines of newsprint joined up. Similarly, the geological 'newsprint' reads straight across the join when the continents are 'put back together' in reconstructions.

In spite of this evidence the idea of continental drift remained a minority view for forty years, not least because nobody had a convincing explanation of how the continents could have moved apart. It seemed unlikely (to say the least) that they could have ploughed through the crust of the ocean bed, like great liners ploughing through the sea; and although you might imagine the Earth cracking and the continents being pulled apart if the whole planet was expanding, what on Earth could make it expand in this way – by two-thirds – in about 200 million years? So although the idea was widely discussed in the 1920s and 1930s, it was only in the 1960s that it finally became accepted, because it was only in the 1960s that incontrovertible evidence that the Atlantic Ocean is getting wider was obtained.

In its modern form, continental drift is part of a larger package of ideas that are collectively known as plate tectonics. The whole thing took off as a result of detailed surveying of the crust under the oceans, using seismic techniques, in the 1950s. For these local studies, you don't have to wait for earthquakes to set the Earth ringing like a bell, but can make do with setting off explosions to make sound waves to probe the crust. It was only when they were able to do this on a large scale, far out to sea, that geologists discovered just how thin the oceanic crust is, no more than five to seven kilometres in some places. They also discovered how rugged the sea floor is, with mountains, submarine canyons, and, most important of all, great oceanic ridges that run for thousands of kilometres in length and stand several kilometres above the average height of the ocean floor. The archetypal example is the Mid-Atlantic Ridge, which, as its name suggests, lies roughly halfway between Europe and America, running north–south up the North Atlantic Ocean. Along the centre of this ridge there is an active rift valley, with many sites of underwater volcanic activity dotted along its length.

In 1960 Harry Hess, of Princeton University, explained this and all the other newly discovered features of the sea floor in terms of the first model of what became known as sea-floor spreading (more accurately, ocean-floor spreading, but the alliteration has a nice ring to it), reviving the idea of continental drift in a new context. According to this model, the ocean ridges are produced by upwelling convection currents in the fluid material of the Earth's mantle.[2] These slow convection currents carry material up to the surface of the Earth at the ocean ridge, where it spreads outward on either side, pushing the continents apart and forming new, young ocean basins from the material forced out from beneath the surface.

Convection, of course, involves material rising up in one place and sinking somewhere else, and in order for the Earth to stay roughly the same size throughout recent geological history crust must be destroyed elsewhere at more or less the same rate that it

[2] Even though the mantle is solid, because it is hot the material can still move in very slow convection patterns, flowing rather like the material known as silly putty. It is *fluid* in this sense, but not *liquid*.

is being squeezed out from oceanic ridges. Hess pointed the finger at the deep trench systems which mark the edge of some oceans, especially the western Pacific. There, he suggested, thin oceanic crust is being dragged back down, under the edge of the continent, eventually to melt into the asthenosphere and close the convective loop. The model also offered an explanation for all of the geological activity going on along the western Pacific, where, as anyone who has visited Japan knows, volcanoes and earthquakes are common. On this picture, the Atlantic is getting wider, at a rate of about two centimetres a year, but the Pacific is shrinking, with North America slowly moving towards Asia.

Even at the beginning of the 1960s the idea of sea-floor spreading did not immediately sweep all before it – it was not one of those 'Eureka!' moments when everyone says, 'Of course, why didn't I think of that?' The persuasive evidence came in over the next few years, when geologists studied the sea floor using a new technique to probe the magnetic properties of the rocks.

Rock magnetism is a key tool for geology on land. The Earth's magnetic field is not constant, but changes over geological time, sometimes weakening and then getting stronger again in the same sense, sometimes reversing completely, so that what is now the north magnetic pole has in the past sometimes been the south magnetic pole. This process is only poorly understood, although certainly it has something to do with the way the electrically conducting material of the liquid outer core swirls around deep inside the Earth. Unfortunately, the term polar reversal is often misunderstood, and alarmists sometimes talk of the Earth toppling over in space, or the whole crust suddenly slipping around the globe (while the magnetism stays as it is), so that Australia and Europe change places. This does not happen! All we are talking about is a change in the internal dynamo of the Earth so that the magnetic field fades out and comes back in the opposite sense, while the continents stay where they are, on the short timescale this takes.

On land, when rocks are laid down (for example, by magma flooding out of a volcano and setting), the strata start out laid one on top of the other. As a layer of molten rock sets it becomes magnetised, picking up magnetism from the Earth's magnetic field, so that its natural magnetism is in the direction that the Earth's magnetic field has at the time the rock is laid down. In well-preserved layers of strata the changes in the magnetism of the Earth can be seen by counting down the strata and noting the reversals of the magnetic orientation of the rocks.

Often, though, the strata have been twisted and jumbled up by geological processes, such as mountain-building. In such cases, once the magnetic reversals have been calibrated in the well-preserved strata the technique can be reversed. To some extent, the correct ordering and dating of the jumbled strata can be reconstructed by looking at their magnetic properties, and matching these up to the magnetic properties of the well-ordered strata.

When magnetometers were first towed behind ships at sea to measure the magnetism of the rocks of the ocean floor, though, they revealed a very different, but striking, pattern. The rocks of the Atlantic sea floor carry a fossil magnetism which runs in north–south stripes, with one stripe having the same orientation as the Earth's magnetic field today, the next stripe having the opposite orientation, and so on. What's more, the pattern of these stripes on one side of the Mid-Atlantic Ridge is the mirror image of the pattern on the other side of the ridge.

The conclusion is inescapable – the magnetic stripes correspond to the orientation of the Earth's magnetic field at the time when the rocks were solidifying as they squeezed out evenly on either side of the ridge, like toothpaste being squeezed from a tube. For millions of years rock was squeezed out and picked up one kind of magnetism. Then, in the blink of a geological eye (a few thousand years), the field reversed, and for the next interval of millions of years the rock being squeezed out picked up the opposite magnetic orientation. This repeating pattern has left the rocks of the sea floor with a record of the Earth's changing magnetism almost literally like a tape recording, with the youngest rock next to the ridge and the oldest on either side of the ocean. It was, in fact, the discovery of this distinctive pattern of magnetism on the sea floor which first revealed the way in which the Earth's magnetic field has reversed repeatedly during the recent history of our planet. But since the pioneering days of the

1960s the magnetic stripes on the ocean floor have been matched up with the magnetic pattern seen in vertical layers of strata on dry land. Everything fits together beautifully.

With all this new evidence coming in, the time was ripe for continental drift, in its new guise of plate tectonics, to be accepted. One of the key moments, when a majority of experts began to swing in support of the idea, came at a meeting organised by the Royal Society in 1964, where Edward Bullard, of the University of Cambridge, presented one of the first reconstructions of the fit of the continents on either side of the Atlantic made by an electronic computer. Strictly speaking, this was no more (and no less) persuasive than Snider's hand-drawn reconstruction had been in 1858, except that it used the edge of the continental shelf to define the edge of each land mass instead of using the present-day shoreline as a guide; but computers still had a certain mystique in 1964, and the ground had been prepared by all the work on sea-floor spreading. Bullard caught the wave, and captured the imagination of his colleagues, so that 1964 is widely seen as the year when the idea of moving continents came in from the cold. Any lingering doubts were dispelled in the 1970s, when laser range-finding using satellites became accurate enough to measure the movement of the continents, and confirmed that, among other things, the North Atlantic really is getting wider at a rate of a couple of centimetres a year.

The term 'plate tectonics' itself was first used just three years after Bullard presented his map to the Royal Society meeting, in a paper in the scientific journal *Nature* in which Dan McKenzie and R. L. Parker brought together all the new ideas in geophysics in one coherent package. By the end of the 1960s the model was essentially complete. It explains how the bulk of the Earth's crust (oceanic and continental) is made up of a few plates, with relatively little geological activity going on in the middle of the plates. The plates fit together like the pieces of a jigsaw puzzle to cover the entire surface of the Earth; but, unlike a jigsaw puzzle, activity at the boundaries between the plates changes the pattern as time passes. There are only six main plates, and twelve smaller plates, that together cover the entire globe.

The way the pattern of the plates is changing depends on what goes on at the boundaries between plates – the margins, as they are known. There are three types of margin. Constructive margins are places where new oceanic crust is being created at ocean ridges and spreading away on either side, so that we see two plates moving away from one another. Destructive margins occur at the deep trenches, where the thin crust of an ocean floor slides under the edge of a thick slab of continental crust, diving at an angle of about 45° down into the mantle, so that we see two plates moving towards one another and one being annihilated. Finally, in conservative margins plates are neither created nor destroyed, but slide past one another.

Individual plates may be made up solely of continental crust, or solely of oceanic crust, or a mixture of both; but constructive and destructive margins can only occur in association with oceanic crust. Continental material today is neither created nor destroyed by tectonic processes, except (possibly) for a little mountain-building that goes on along the continental side of a destructive margin, where the volcanoes are active. Nobody can say for sure how the continental crust came into existence in the first place – what formed the first chunks of continental crust which could get involved in mountain-building in this way. The most likely explanation, though, is that the young Earth was scarred by the impacts of large asteroids, and that the first mini-continents were the scabs that formed over these wounds. Whatever its origins, if continental crust is carried up to a destructive margin by a plate which is being swallowed up there, it cannot be destroyed by the trench, and the destructive margin soon ceases to operate.

Overall, oceanic crust is being created at spreading ridges at just the same rate that it is being swallowed up in the deep trenches. If a deep trench gets blocked up with continental material, the repercussions are felt around the tectonic conveyor belt, and a spreading process somewhere else in the world must slow down or stop in compensation. The whole pattern of tectonic activity is a variable feature of the Earth.

The best way to explain all this is with examples. The Red Sea and the East African Rift Valley are part of a complex system of faults in the Earth's crust, where spreading activity has begun relatively recently, with hot material welling up from below the crust and spreading sideways by convection, cracking the crust as it does so. The Red Sea looks like an ocean in miniature, and even has a central 'spine' formed by a spreading ridge. Although it is too early to be sure exactly which of the faults in east Africa will develop fully in the geological future, it does seem likely that over tens of millions of years one of these rifts will combine with the Red Sea and develop into an ocean as big as the Atlantic, splitting the bulk of Africa away from Asia.

In western North America, by contrast, the continent, moving westward as a result of the spreading of the North Atlantic, has overrun a former spreading ridge which is now barely active, although traces of its former power can still be felt. This happened because the spreading activity in the North Atlantic was going on faster than the spreading at the equivalent ridge in what is now the North Pacific. The American continent was being pushed westward faster than the Pacific ridge could create new crust, so for a long time the Pacific Ocean crust was disappearing down a deep trench along the west coast of North America, being gobbled up faster than it was being produced at the spreading ridge. It is as if an escalator, with a moving floor being produced steadily at the bottom, was being 'eaten' twice as fast at the top, so that the top edge of the machine advanced steadily downwards.

The Pacific spreading ridge is still active further south, and it is like a knife sticking into California, where it is responsible for the splitting away of the Baja California peninsula, and for the activity of the notorious San Andreas fault. By and large, though, the boundary between the North American plate (which includes the western half of the North Atlantic ocean floor) and the Pacific plate (which is essentially the entire Pacific Ocean sea floor) is a conservative margin, with the Pacific plate twisting slightly and rubbing northwards past the continental plate. In the relatively recent geological past the activity associated with this collision between plates has thrown up the Rocky Mountains; today, it is responsible for the earthquake and volcanic activity up this entire seaboard (and, indeed, similar plate boundary interactions explain the seismic activity right around the Pacific Rim, which is sometimes described as 'the ring of fire').

This activity in California is associated with a northward movement – at a rate of about six centimetres a year – of the crust to the west of the San Andreas fault, which includes a sliver of the continent that has been split away from the rest of North America by the last trace of activity on the old spreading ridge. What is now San Francisco was more than a thousand kilometres further south 30 million years ago, when the ridge and trench systems began to annihilate one another (if, and it is a big *if*, the process has gone on steadily for all that time). But this northward movement is not steady; in some places, the two plates stick together for decades at a time, sometimes for centuries, then jerk forward, releasing the accumulated tension. Which is why southern California is prone to large earthquakes – if the region around Los Angeles, say, 'sticks' for a hundred years, when it does move there is a full six metres of jerk required to catch up with where the plates belong. And a sideways shift of six metres makes for a very big earthquake.

There are also extinct faults in many parts of the world. The Great Glen in Scotland, for example, marks the site where there used to be a conservative plate boundary very similar to the one that now exists in California, but, in this case, with two pieces of continental crust sliding past one another. The activity along this boundary has long since faded away (although even Scotland does experience minor tremors from time to time), and the two plates have become welded together; but the boundary between two formerly distinct continents is still marked by the rift of the Glen itself.

There are also more complicated possibilities, such as the tectonic processes that occur at triple junctions, where three plates meet, or where a spreading ridge meets a destructive trench at right angles. But we need not go into such details here. An individual plate can be any shape, and can be bounded by any combination of constructive, destructive or conservative margins, as long as the whole global picture adds up to keep the total amount of crust constant. As the Pacific plate contains no continental crust, and it is being destroyed at its western margin, it is quite possible (even likely) that one day it will be destroyed altogether, as North America and Asia collide to form a new supercontinent. By then the Atlantic may have widened to the size of the Pacific Ocean today, and Africa may be separated from Arabia by an ocean the size of the present-day North Atlantic.

Such collisions and rearrangements of the continents, followed by a break-up of the supercontinent into smaller pieces once again, as new spreading sites develop, seem to have occurred several times in the history of the Earth. When continents collide, mountain chains are thrown up. The Himalayas, still growing upward today, are the direct result of India ploughing northward into Eurasia; the mountains of Cyprus are being squeezed upwards as the European and African continents move together, crushing the Mediterranean Sea between them. Mountains also form, as we have mentioned, along the edges of destructive margins. So, wherever there are mountain chains far from any plate boundaries today, we can be sure that they mark the sites of long-extinct plate boundaries. By 'tearing along the dotted line' of a mountain chain and pulling the continents away from one another along the join in their reconstructions, geophysicists can get some idea of how the continents used to look long ago. The older the mountain chains involved, the longer it has been since that boundary was geologically active.

There is also other information about the positions of continents in the past. The fossil record of life reveals whether the rock in which it is found was near the tropics or in colder climes. Scouring by glaciers during ice ages leaves scars that can be seen on continents that are far apart today, but must have been joined in one land mass when the glaciers lay over them. The magnetic record also shows how the rocks have moved, because magnetism is always laid down running north–south at the time the rocks cooled, but may be found today pointing in any direction, showing that the whole continent in which the rocks are found today has twisted around as it has moved. In this way, using a variety of geological evidence, it has been possible to sketch out, at least in outline, how the geography of our planet has changed over the past 600 million years or so – although the outline is, of course, more sketchy the further back we look, and very little can be said about the geography of even earlier times.

Geological evidence shows that more than 600 million years ago most of the land that now forms South America, Africa, the peninsula of India, Antarctica and Australia was grouped together in one supercontinent, which is known as Gondwanaland. To put this in some sort of perspective, 600 million years ago was a hundred million years before the evolution of the first fishes; at that time, Gondwanaland lay roughly across the equator. It drifted slowly southwards as a block (thanks to the activity of some long-gone spreading ridge or ridges), and moved over the South Pole around 450 million years ago, triggering the formation of glaciers which have left their scars across all of the modern-day fragments of Gondwanaland.

Meanwhile, near the equator and in low northern latitudes other pieces of the continental jigsaw puzzle were assembling themselves. North America and Greenland had, by that time, been attached to each other for hundreds of millions of years. By about 400 million years ago, at the time when land plants were beginning to spread over the continents, this chunk of continental material collided with what is now part of Europe (notably Scotland and Scandinavia), and the pieces welded together to form what is known as the Old Red Sandstone continent, named after rocks that were formed during the collision. By then Gondwanaland had crossed the South Pole and was moving northward, while vertebrates were moving on to the land, and the first reptiles evolved. A little over 250 million years ago, Gondwanaland and the Old Red Sandstone continent collided and stuck together. Shortly afterwards, the last remaining independent large chunk of continental crust, present-day Asia, collided with the northern part of this supercontinent and was welded on to Europe, throwing up the Ural mountains in the process. All of the modern continents were joined together in one supercontinent, known today as Pangea, stretching from the South Pole right across the equator to high northern latitudes.

We know much more about the way tectonic activity has changed the geography of the globe since the time of Pangea than we do about earlier times, because a lot of the evidence of earlier tectonic activity was destroyed, or at least blurred, by the collisions that formed Pangea. There were glaciers at both the northern and the southern extremes of Pangea, and the circulation of the oceans was severely restricted by the way the supercontinent stretched from north to south, stopping the ocean currents from travelling right around the world. At that time, known as the late Permian period of geological time, many forms of life on Earth died out, probably as a result of the cold and the way the sea level fell as water became locked up in the ice caps. The crisis for life was so dramatic that geologists have literally made it the end of an era, the Paleozoic.

All the intervals in the geological timescale are defined in this way, in terms of changes in the fossil record. Eras are long intervals whose boundaries are marked by very dramatic changes in the forms of life around on Earth, and periods are shorter intervals (subdivisions of eras) marked by lesser extinctions of life – except when the end of an era and the end of a period coincide. Starting around 250 million years ago, in the era that followed the Paleozoic – the Mesozoic – life recovered as Pangea began to break up and the world warmed. The earliest mammals actually appeared early in the Mesozoic; but for hundreds of millions of years they took a back seat, as the era became dominated (at least, as far as land animals were concerned) by the dinosaurs.

Pangea had, indeed, begun to break up almost as soon as it had formed. The first great rift occurred along a line that corresponds to the present-day Mediterranean and the Caribbean, with more or less the old Gondwanaland breaking free of the northern continents once again, and then breaking up into the pieces that we know today. In the north, the supercontinent of Laurasia, left behind by the break-up of Pangea, also began to break apart into smaller pieces. The major rift then was between southern North America and Europe, with Greenland also – eventually – splitting from North America and getting left behind as the North American continent moved westward. The opening of the North Atlantic was one of the last acts in this break-up, and northern North America, Greenland and Europe were still joined in the north about 65 million years ago, at the time of the death of the dinosaurs. Coincidentally, this last great break-up of the northern supercontinent occurred at about the same time as Australia broke free from Antarctica in the last major rift of what was left of Gondwanaland. Since then, almost exactly coinciding with the era of the mammals, the story of continental drift has largely involved the continents as we know them today, moving slowly into the positions that they occupy now.

It has taken all of the past 65 million years or so for the North Atlantic to widen to its present size, and this interval, since the death of the dinosaurs, marks another era (as yet incomplete) in geological time. It is the Cenozoic, the era of the mammals. Compared with some of the numbers we have just been tossing around, 65 million years sounds like a short interval – indeed, by geological standards, it *is* a short interval. But during the Cenozoic, the era of the mammals, half of all of the sea floor of the Earth has been pushed down into deep ocean trenches by the sea-floor conveyor belt and destroyed, replaced by new oceanic crust spreading out from ocean ridges. The oceans cover two-thirds of the surface of the Earth; so this means that, since the time of the dinosaurs, one-third of the entire surface crust of the Earth has been recycled and renewed.

It is the tectonic activity of the Earth today which makes some regions more prone to earthquake and volcanic activity than others. Seismic activity commonly occurs at plate margins, of all varieties, with volcanic activity being particularly common where oceanic crust is being forced down deep trenches and melting, producing plumes of hot, molten rock which rise up to burst through the mountain ranges that form on the nearby continent, where continental crust is being crumpled up as the continent overrides the trench.[3] As the fate of Pangea and Gondwanaland shows, nowhere on Earth is safe from seismic activity in the long run. In the very long term new rifts may open up

[3] The volcanic islands of the Pacific, such as Hawaii, are an exception to this rule. Far from any plate boundaries, these are sites where a hot spot in the mantle has punched a hole in the thin oceanic crust. As the Pacific plate drifts across the hot spot, it actually punches a series of holes in the crust, producing a chain of volcanic islands. Intriguingly, all of the Pacific island chains of this kind have a bend in them, showing that the drift of the whole plate changed direction, from a more northerly drift to the northwesterly drift we see today, about 40 million years ago. That was when North America overran the northward part of the old Pacific spreading ridge, giving the plate a westward nudge as it did so.

anywhere, even in the middle of a continent, tearing it apart the way Africa is being torn apart along the Great Rift Valley system today. And even places like the Highlands of Scotland, or the Appalachians, or the Ural mountains (which are all far from any plate boundaries today) may still experience rumbles of activity as the old joins between former continents settle a little more firmly into place.

All of this activity is directly relevant to the story of life on Earth, and to our central theme, the relationship between humankind and the Universe. Life has been around on Earth almost from the moment that the planet cooled – traces of bacteria have been found, as fossilised remains, in rocks more than three billion years old. Just how life first got a grip on the planet is not clear, but geological activity may have played an important part here, as well. The old idea is that a mixture of chemicals brewing up in a shallow sea (or what Darwin referred to as 'a warm little pond'), energised by sunlight and violent electrical storms, may have produced the first self-replicating molecules. A minority (but respectable) view is that the first replicators formed somewhere else in the Universe, perhaps in clouds of material in space, and were brought to Earth by comets. The merit of this idea is that it means there were billions of years before the Earth formed in which the necessary prebiotic processes could have gone on, explaining why life got a grip on Earth so quickly after the planet formed – it didn't have to start from scratch. But one of the latest suggestions builds from the discovery of weird life forms that live far beneath the surface of the oceans today, getting their energy not from sunlight but from the heat that emerges from volcanic vents, cracks in the sea floor, producing boiling plumes of water and an active chemistry involving hydrogen sulphide. Just possibly, this sort of activity might have produced the first self-replicating molecules on our planet.

We know very little about the early history of life on Earth for two reasons. First, the life forms were soft-bodied, and had no skeletons or other hard parts that would easily form fossils when they died. Secondly, at first life existed entirely in the sea, so that any remains that were left were on the sea floor, which has been recycled completely several times by tectonic activity since then – except where that activity has lifted up the sea floor and made it part of a continent, as is happening in the Mediterranean today.

The boundaries of geological time are marked by changes in the fossil record of life on Earth, and the greatest of these boundaries is the one between the Precambrian era, which represents the entire history of the Earth before 600 million years ago, and everything more recent. For almost four billion years evolution proceeded in the oceans in the form of softbodied, single-celled creatures that have left us few remains to study. Towards the end of the Precambrian, though, life began to diversify into multicellular forms, with jellyfish, worms, sponges and the like appearing in the (still sparse) fossil record. Then came the Cambrian explosion – not so much an explosion of life as an explosion in the fossil record – after about 600 million years ago when living creatures first developed features such as shells which were easily fossilised and (equally important) are easily recognisable today.[4] But this was still long before life had emerged on to the land, where the continents were nothing more than barren rock. Although life did eventually transform the land, the rocks themselves have also played a part in the evolution of life, as their slow drift around the surface of the Earth has brought marked changes in the environment in which life operates. And this happened even before life moved onto the land.

About 440 million years ago there was a massive extinction of life on Earth, with many species dying out. The catastrophe is so obvious in the fossil record that it is used to mark the end of a geological period, the Ordovician. In that fossil record, this seems to be the second-greatest extinction that has ever affected life on our planet. It happened just at the time when Gondwanaland, which had formed during the late Ordovician, drifted over the South Pole and became partly covered in ice.

When an ice age is triggered in this way it isn't just the polar regions that are affected. Ice can only grow over the poles of our planet if the supply of ocean currents carrying warm water from

[4] The Cambrian, by the way, is not another era, but merely a geological period, which lasted about a hundred million years. It is the first period of the Paleozoic era.

tropical latitudes is cut off, and the most straightforward way for this to happen is to have a land mass sitting astride one of the poles.[5] The land also provides a surface on which snowfall can settle. Instead of falling into the sea and melting, it can build up, forming ice sheets and glaciers as the centuries pass. The ice and snow themselves then contribute to the cooling, because they reflect away heat from the Sun that would otherwise warm the surface of the Earth, be it land or sea. The result is that the whole planet, not just the polar region affected directly, feels the chill, and the patterns of wind and weather change around the globe.

For life on land, as we shall see later, a sometimes critical side-effect is that because the whole planet is cooler there is less evaporation from the oceans, so the winds are drier and less moisture is available in the air to fall as rain over the land. This didn't matter much to life at the end of the Ordovician, because there was no life on the land. What we do see clearly in the fossil record of the marine organisms from that time, though, is that species that were adapted to cold-water conditions moved towards the equator. But species that were adapted to warm water conditions – the majority – had nowhere to go. They died.

It was after this disaster that life not only recovered but moved on to the land, and continental drift played a part in that process as well. As Gondwanaland moved away from the South Pole the burden of ice it carried melted. Warm water returned to high southern latitudes, and the ice age ended. At the same time – or rather, over a few tens of millions of years – the North Pole was covered by oceans, and the fragments of land that today form North America, Europe and Asia were scattered around the equator, with shallow seas between them and washing their shores. The interval from about 440 million years ago to about 360 million years ago (which encompasses two geological periods, the Silurian and the Devonian) was particularly warm, perhaps because the volcanic activity associated with these continental movements had made the atmosphere rich in carbon dioxide, a gas which traps heat from the Sun through the so-called greenhouse effect, and this encouraged life to flourish. But

[5] Or both, but it was only the South Pole that was affected at the end of the Ordovician.

the most important feature of the environment then (from our own perspective) may have been the shallowness of the seas that edged the land masses at that time. This meant that there were large areas that were entirely covered by water at high tide, and left uncovered at low tide.

The extra carbon dioxide that warmed the world in the Silurian also encouraged plant growth – plants absorb carbon dioxide during photosynthesis, and use it to build their tissues. As Carl Sagan put it, in one of his most memorable phrases, 'a tree is largely made of air'. Many plants grow more strongly when the supply of carbon dioxide increases. Plants growing in shallow tidal estuaries and river deltas in the Silurian would have been uncovered at low tide, so there would be a risk of them drying out; but they would also receive more direct sunlight, the other essential ingredient for photosynthesis. In the struggle for survival, plants that could tolerate drier conditions would have had a clear advantage, moving out of the crowded tidal waters on to dry land, and obtaining uninterrupted supplies of sunlight.

Animal life soon followed the first plants on to the land, for the obvious reason that the seas were crowded and land, once there were plants there to provide food, offered something of an escape route. It's worth stopping for a moment to think this through. The most successful forms of life are the ones that are superbly adapted to their ecological niches, and never have to change much. Many bacteria that live on Earth today are essentially identical to their ancestors that were around when the world was young, three billion years ago. They are the great survivors, the true successes of evolution.

But when individuals are finding it tough going in the competition for food and other resources, they have to, literally, find pastures new.[6] The most successful fish stayed as fish, and their descendants are still fish today. It was the fish that weren't doing too well in the seas that were pushed to the margins, the tidal mudflats where they developed into amphibians, moving inland

[6] This only works if the new pastures are not already occupied. The reason life in the sea doesn't move on to the land today is that there is plenty of land-based life around already, and any hypothetical proto-amphibian that flopped out of the water would soon get eaten.

to eat the insects which had followed the plants out of the shallow seas. Similarly, it was the *least* successful amphibians that had to move on, adapting completely to dry land and becoming reptiles, and so on. We are descended from a long line of creatures that were not very good at their roles, and had to adapt or die – a long line of near failures, in evolutionary terms (not *complete* failures or they would have left no descendants at all). Even worse, in terms of putting us in our place, our lineage was not the first to move on to the land. It was segmented sea-dwellers with hard shells who found it easiest to move out of the sea, once food was available. Ancestors to millipedes were among the first animals to 'conquer' the land; and cockroaches were well established on land 300 million years ago, while our ancestors had scarcely stopped splashing about in the shallows.

In fact, our direct ancestors, the vertebrates, moved on to the land just after another great extinction, which marked the end of the Devonian period, about 360 million years ago. Once again, the drift of Gondwanaland was largely to blame for the extinctions, twisting the continent and moving it back over the South Pole. In many ways, this late Devonian crisis for life was a replay of what had happened at the end of the Ordovician. It particularly affected life in the sea, with life forms adapted to cool temperatures moving towards the equator, and life forms that were adapted to the tropics being wiped out. There is also evidence that the Earth may have been struck by a large object from space, an asteroid or a comet, during the ice age associated with the drift of Gondwanaland back over the South Pole.

The main effect of such an impact would be to spread a shroud of dust high in the atmosphere (fine dust, like talcum powder), reflecting away incoming solar heat and cooling the planet still further. Whatever the exact causes of the cooling, though, there is no doubt that the Earth once again went into deep freeze at the end of the Devonian, many forms of life went extinct as a result, and that it was in the recovery after this great extinction, with the survivors radiating out to fill the vacated ecological niches, that the amphibians emerged out of the seas and on to the land.

It was the beginning of the Carboniferous period, the time

when huge forests covered large areas of swampy, low-lying land. When the trees in those forests died and fell, their remains, rich in carbon stored by photosynthesis, were preserved in the swamps, where they built up into thick layers that were then squeezed by geological forces and, over a long interval of time, turned into coal. This period lasted from 360 million years ago to about 286 million years ago, and is sometimes subdivided into the Mississippian (from 360 to 320 million years ago) and the Pennsylvanian (from 320 million years ago to 286 million years ago), named after coal deposits with roughly those ages that are found in the appropriate parts of North America today.

Much of the coal that is burned around the world today, and is largely responsible for the increase in the greenhouse effect that we are now experiencing, comes from deposits laid down in the Carboniferous. Today, this poses a problem of global warming, as we discuss in the next chapter. But so much carbon dioxide was taken out of the air by plants and locked up as coal in this way in the Carboniferous that it must have had the opposite effect, cooling the planet. Frustratingly, we don't know exactly what effect this may have had on life, since we don't know how much carbon dioxide was in the air at the start of the Carboniferous, and how much the greenhouse effect was weakened.

By then, though, the evolutionary action, as far as the lineage that leads to us is concerned, had moved on to land for good. The Carboniferous was followed by the Permian period, which lasted until 248 million years ago, and it was about halfway through the Permian, some 270 million years ago, that the first warm-blooded animals arrived on the stage – proto-mammals. Throughout the late Permian, almost all of the continental crust of the Earth was joined in the single supercontinent of Pangea, which stretched from pole to pole, and which caused another interval of cold. Warm-blooded creatures were much more successful than their cold-blooded rivals under such conditions. Just when they seemed to be flourishing, though, life on Earth was struck by a disaster which brought the biggest extinction in the entire fossil record. It was so spectacular that geologists use it to mark not just the end of a period (the Permian) but also the end of an era (the Paleozoic).

In the seas, the extinctions at the end of the Permian (the 'Permian terminal extinctions') killed off 90 per cent of all species, over an interval of at most 10 million years, perhaps much less. On land, it is harder to tell what proportion of species went extinct, but this was the first, and greatest, extinction to affect land-based vertebrates, and was particularly severe on the proto-mammals, including our direct ancestors. The cooling of the globe associated with the build-up of Pangea must have made life difficult for millions of years; but it is hard to escape the conclusion that some additional catastrophe, almost certainly a blow from space, tipped an already difficult situation into a disaster for life on the land.

During the Triassic period, which followed the Permian and lasted from 248 million years ago to 213 million years ago, it was the reptiles, not the mammals, who recovered best from the crisis and spread to fill the available ecological niches. By about 230 million years ago, dinosaurs had emerged on to the evolutionary stage. They were so successful (partly because they were probably also hot-blooded) that they dominated two entire geological intervals, the Jurassic (from 213 million years ago to 144 million years ago) and the Cretaceous (from 144 million years ago to 65 million years ago). In round terms, the 'day' of the dinosaur lasted for 150 million years (for comparison, by the most generous definition of the term 'human', we have been around for 5 million years). During the time of dinosaur domination, Pangea broke apart, first forming two separate supercontinents, Laurasia and Gondwanaland II, and then with those supercontinents in their turn breaking up and the fragments drifting towards the positions they occupy today. But no continents drifted over the poles, so there were no great ice ages; and there were no great continental collisions. This was a long interval of relative geological stability, which is why the Jurassic and Cretaceous periods are such long intervals in the geological record.

The extinctions that mark the boundary between the Jurassic and the Cretaceous come in the middle of this long interval of geological calm, and are something of a puzzle, in terms of continental drift and plate tectonics, because there is no obvious reason why tectonic activity should have caused any great changes in the environment then. Both marine creatures and the dinosaurs (and other land-based creatures) certainly did suffer seriously at the end of the Jurassic, and this is one of the prime candidates for a disaster triggered mainly, if not entirely, by an impact from space. But although many species of dinosaur were wiped out and new ones took their places, the dinosaur line continued to dominate the land fauna. They survived in a changing world by adapting and evolving. Among the changes they had to adapt to was the spread of grasses and flowering plants across the land, which happened in the middle of the Cretaceous, a little over a hundred million years ago. By then, dinosaurs had already been around for well over a hundred million years themselves; the interval from the spread of grasslands back to the origin of the dinosaurs is as great as the interval between ourselves and the spread of grasslands. And none of the Jurassic dinosaurs (including such childhood favourites as Diplodocus and Brontosaurus) ever saw a flower or walked on grass.

Even the dinosaurs were unable to survive the catastrophe that struck the Earth at the end of the Cretaceous, 65 million years ago. Some of the changes that brought new environmental pressures to bear at that time were, once again, linked with changing geography. Broadly speaking, on both sides of the proto-Atlantic Ocean, at the end of the Cretaceous there were continents in the south (South America, Africa, and India in particular) and continents in the north (notably North Africa, Europe, and Eurasia) separated by a stretch of open ocean, the Tethyan Seaway, which ran right around the world at low latitudes. Much of this open water was squeezed out of existence as the continents moved together (the Mediterranean is a vestige of the Tethyan Seaway), changing ocean circulation patterns, altering the climate, and being particularly bad news for marine creatures that inhabited the shallow sea itself.

But there is now compelling evidence that the catastrophe for life on land at the end of the Cretaceous, and especially for the dinosaurs themselves, was triggered by the impact of an object from space, striking the Earth in the region of what is now the

Yucatan peninsula of Mexico. The crater formed by the impact has been identified, and was caused by the impact of an object more than ten kilometres across, travelling ten times faster than a rifle bullet, and releasing energy equivalent to at least a hundred million megatons of TNT as it was brought to a halt and its energy of motion was converted into heat.[7]

Material that was blasted out of the solid Earth by the impact would have been hurled up into space and fallen down again, and everywhere the fragments struck, their energy of motion would be turned into heat. Many of the creatures on the land surface of the Earth – including many dinosaurs – were roasted alive as a result, while the heat triggered a global forest fire that burned so fiercely that it has left a layer of soot still visible in rock strata 65 million years old. After the heat came the cold, as smoke and dust soon shrouded the planet and stopped the heat from the Sun reaching the ground. Plant life, dependent on sunlight for photosynthesis, suffered severely; plant-eating animals had nothing to eat. And as the vegetarians died out, the meat-eaters in their turn starved. Altogether 70 per cent of all species on Earth were wiped away in the terminal Cretaceous event, a disaster so severe that it is chosen by geologists to indicate not just the end of the Cretaceous period but the end of the Mesozoic era, and the beginning of the Cenozoic era.

Among the survivors there were some small mammals, shrew-like creatures that had lived for tens of millions of years in the shadow of the dinosaurs – our ancestors. The way they radiated and adapted after the extinctions is another perfect example of Darwinian evolution at work, and it explains why we are here today.

During the geological periods that followed the death of the dinosaurs (the Paleogene, from 65 million years ago to 24 million years ago, and the Neogene, which started 24 million years ago and has not yet ended), the geography of the globe has changed only slightly, although those changes may have played an important part in our own evolution. It is time to leave the story

[7] The compelling evidence that the terminal Cretaceous event was, indeed, caused by an impact from space is spelled out in our book *Fire on Earth* (Simon and Schuster).

of the way the solid Earth has changed, and look at the much more rapid changes which go on in the blanket of air around our planet, and which have consequences relevant to human activities past, present and future.

CHAPTER EIGHT

WINDS OF CHANGE

The atmosphere of the Earth is essential for our well-being, and without it there would probably be no life on Earth at all. Yet in planetary terms it is scarcely more than a geological afterthought. The diameter of the solid Earth is 12,756 km. Defining the exact thickness of the atmosphere is tricky, because it thins out and fades away into space, rather than having a sharp edge, but the altitude above sea level of the top of Mount Everest, at 8.85 km, is just about as high as a person could go and still find breathable air, while the top of the layer of the atmosphere in which weather occurs (the troposphere) is, on average, about fifteen kilometres above sea level. If our planet were the size of a basketball, the thickness of the breathable atmosphere would be no more than one quarter of a millimetre, a barely noticeable smear over the surface of the ball. Or look at it another way. In terms of travelling about on the surface of the Earth, a distance of ten kilometres (just over six miles) is an extremely small journey. Starting from most places on Earth, if you travelled ten kilometres in any direction over the surface you would expect to find the environment much the same as it was where you started. However, if you travelled ten kilometres vertically upwards and tried to breathe the air there, you would die.

And yet, as we have said, this thin smear of air over the surface of the Earth is absolutely crucial in making our planet a suitable home for life. It all began right in the beginning, just after the Earth formed. We don't know a great deal about conditions on the surface of the Earth then, but we do know, from the geological evidence, that sedimentary rocks were being laid down 3.8 billion years ago, and there are traces of life in those rocks from 3.5 billion years ago. Sedimentary rocks are only laid down under water, so there was definitely liquid water in abundance over the surface of the Earth 3.8 billion years ago, less than a billion years after the planet formed. This is intriguing, because astronomers calculate, from their models of how stars work (*see* chapter ten), that at that time the Sun was producing only about 75 per cent of the heat that it produces today. Other things being equal, that would have meant that the early Earth was a lifeless ball of ice, with a surface temperature well below the freezing point of water. And since ice is a good reflector, even though the Sun warmed up as it grew older, a frozen early Earth might well have stayed frozen for ever, reflecting away the extra heat instead of absorbing it and using it to melt the ice.

So why was the early Earth warm enough to have oceans of liquid water in which life evolved? Almost certainly, it was thanks to the greenhouse effect. The greenhouse effect is the process by which atmospheres keep planets warmer than they would otherwise be. It gets its name because the air in a greenhouse is also warmer than it would be if the greenhouse were not there – confusingly, though, a greenhouse does the trick in a different way. An actual greenhouse traps heat because the glass roof acts as a physical barrier. Hot air near the ground, warmed because the ground itself (and all the things in the greenhouse) is warmed by the incoming rays of the Sun, cannot rise by convection and escape because the roof is in the way. The 'greenhouse effect' works in quite a different way. Energy from the Sun, mostly in the form of visible light, passes through the air almost unimpeded, and warms the surface of the land or sea. The warm surface also radiates energy, but at longer wavelengths (because it is cooler than the surface of the Sun), in the infrared part of the spectrum. This infrared radiation is largely trapped by gases in the atmosphere, such as carbon dioxide and water vapour. As a result, the bottom of the atmosphere gets warmer, triggering convection and setting all the weather systems of the world in motion.

Just after the Earth formed it was an airless ball of molten rock. But as the rock cooled and a crust formed, gases escaping from volcanoes and cracks in the crust began to form an atmosphere. These gases must have been the same sort of mixture that emerges from volcanoes today. They included water vapour, which fell as rain and formed the oceans, nitrogen, and ammonia, which was split apart by sunlight into hydrogen and nitrogen. Most of the hydrogen escaped into space, because its molecules are so light; nitrogen is a stable, unreactive and relatively heavy gas, so the primordial nitrogen built up in the atmosphere, where it makes up 78 per cent of the air that we breathe today. But there was also an enormous amount of carbon dioxide in the early atmosphere of the Earth, and carbon dioxide is a very effective greenhouse gas.

We know how much carbon dioxide there was, because it has been preserved in the form of carbonate rocks, which were laid down by processes associated with carbon dioxide from the air dissolving in water – the stalactites and stalagmites in limestone caves are examples of this process at work. The pressure exerted by the atmosphere at sea level today is defined as one bar – one atmosphere of pressure. If all the carbonate in all the rocks in the Earth's crust were turned back into carbon dioxide, it would produce a pressure at the surface of 60 bars – sixty atmospheres of pressure. Of course, all of that carbon dioxide was not in the atmosphere at one time. But there is ample to explain how the young Earth was kept warm enough for water to flow, by the greenhouse effect, even though the Sun was cooler long ago. It is the combination of a carbon dioxide atmosphere keeping the young Earth warm, and oceans in which carbon dioxide was dissolved and laid down in rocks, so that the greenhouse effect weakened as the Sun got warmer, which has made the Earth a suitable home for life for more than three billion years. And we can see how things might have been by looking at our nearest neighbours in space – the airless Moon, the planet Venus (next in line to the Earth but nearer the Sun), and the planet Mars (next in line to the Earth but further from the Sun).

One effect of a blanket of air around a planet is to carry heat around the globe, smoothing out temperature differences. On the airless Moon this cannot happen, and the temperature soars to above 100°C on the sunlit side, and falls to around –150°C in the night-time. The average temperature is about –18°C, even though the Moon is at roughly the same distance from the Sun that the Earth is, and its surface warms and reradiates solar energy in much the same way that the surface of the Earth does. By contrast, the average temperature on the surface of the Earth is about 15°C. The Earth is about thirty-three degrees warmer than it would be if it had no blanket of air, and this is entirely due to the greenhouse effect at work – even though the atmosphere today contains only about 0.035 per cent carbon dioxide (and it was only 0.028 per cent a century and a half ago, before people started burning fossil fuels profligately, putting Carboniferous carbon dioxide back into the air). It is straightforward to calculate the size of the greenhouse effect produced by this carbon dioxide and the water vapour in the air, and the calculation, indeed, 'predicts' that the surface would be about thirty-five degrees warmer as a result than it would be if the Earth had no atmosphere – or if the atmosphere was purely made of nitrogen, which is not a greenhouse gas.

So how much greenhouse warming would 60 bars of carbon dioxide produce? This is almost a trick question, because there is only so much infrared radiation going out from the surface of the Earth to be absorbed. A photon that is absorbed by one carbon dioxide molecule won't be absorbed again, however many carbon dioxide molecules there are. Once all the infrared photons are being absorbed in the greenhouse effect, adding more carbon dioxide won't make any difference. Long before we reached 60 bars, there would be a runaway greenhouse effect, with the planet getting as hot as it possibly could.

This is exactly what happened on Venus. Although Venus is almost the twin of the Earth in size, it is closer to the Sun, and receives more heat from the Sun. Even without an atmosphere, its surface temperature would be 87°C, and only a small greenhouse effect would take the temperature above 100°C. Venus started life in just the same way as the Earth did, but this extra solar heating plus the natural greenhouse effect combined to make sure that even when Venus was young the temperature never fell low enough for water to condense out of the atmosphere. All of the carbon dioxide ever outgassed on Venus stayed in the air, because there were no oceans to dissolve it. This added to the greenhouse effect, and temperatures at the surface soared as the Sun warmed. The result is that Venus today has a surface temperature above 500°C, way above the boiling point of water. And how much carbon dioxide is there in its thick atmosphere? In round terms, a little more than 60 bars – the Earth's twin planet has outgassed the same amount of greenhouse gas as the Earth itself has, but the difference is that with no liquid water for it to dissolve in all the carbon dioxide has stayed in the atmosphere. And that is why Venus is a roasting desert while Earth is a suitable place for life.

Mars, which is both smaller than the Earth and further out from the Sun, has only a thin carbon dioxide atmosphere today, and is a cold, lifeless desert. In most parts of Mars, the temperature drops below –140°C every night, and daytime temperatures only climb above freezing for a short period during summer in the southern hemisphere of the planet. But its surface is marked by many features which seem to have been carved by flowing water long ago, when the planet was young. All the evidence suggests that Mars started out along the same path as the Earth, with a greenhouse effect strong enough to raise the temperature at the surface above freezing throughout the year. But because Mars is so small – its mass is only a little more than one-tenth of the mass of the Earth – its gravitational pull was (and is) too weak to hold on to the carbon dioxide, most of which has been lost into space. Once the original atmosphere was gone, it could not be replaced, because Mars is so small. The planet would have lost its internal heat relatively quickly, so that tectonic processes ground to a halt, and there were no active volcanoes to put more carbon dioxide into the atmosphere. So the greenhouse effect on Mars today is only a feeble reminder of what it used to be. Which is why Mars is a frozen desert today, while the Earth is a nice place for life.

But if the nature of the early atmosphere of the Earth had a profound effect in making the planet habitable, once life did get a grip on the planet it began to change the nature of the atmosphere itself. In photosynthesis a series of chemical reactions driven by the energy of sunlight (so they go 'uphill' in the chemical energy sense that we used the term earlier) manufacture complex compounds – carbohydrates – from carbon, hydrogen and oxygen.

The first bacteria that exploited photosynthesis used hydrogen sulphide as their source of hydrogen (one reason why some people think that life on Earth originated around hot volcanic vents) and got their carbon and oxygen from carbon dioxide. Because these bacteria lived in water, they took carbon dioxide out of solution, allowing more carbon dioxide from the air to dissolve, and helping to stabilise the terrestrial environment, with no risk of a runaway greenhouse effect. It was only later that some bacteria evolved that used a different version of photosynthesis, involving chlorophyll, the substance that makes green plants green. In this process the energy of sunlight is used to obtain hydrogen from the water itself, instead of from hydrogen sulphide. Because the organisms get all the oxygen they need for photosynthesis from carbon dioxide, the oxygen left over when hydrogen is removed from water is, to them, a waste product. It is released into the atmosphere, and once this process started it eventually brought about a dramatic change in the composition of the air.

Oxygen is a highly reactive substance, readily combining with many other elements in the process known as oxidation (a burning flame is an example of rapid oxidation; rusting is an example of slow oxidation). At first there were many substances around for the oxygen being released by the new kind of photosynthesisers to combine with, and they seem to have disposed of the dangerously reactive oxygen by locking it up with ions of iron in the water (usually referred to as ferrous ions, from the Latin, rather than iron ions, for obvious reasons). Large quantities of an iron ore known as haematite, an oxide of iron, were laid down on the beds of shallow seas where the early forms of these photosynthesisers lived, between about three billion and two billion years ago, and are mined today in places such as Western Australia, Labrador and the Ukraine.

Then some of the photosynthesisers learned to live with free oxygen, evolving enzymes that protected them from oxidation. This meant they could just throw away oxygen into the environment, without having to take the trouble to lock it away in iron compounds. This was a disaster for many of the forms of life around at the time (roughly two billion years ago), which had no such protection against the reactive power of oxygen, and to whom the gas was a poison. Although the soft bodies of these single-celled creatures have left no trace in the fossil record, one of the greatest extinctions of life must have occurred when the first of this third wave of photosynthesisers learned how to live with free oxygen, with which they polluted the entire world and which had the happy (for them) side-effect of killing off many of their competitors. And oxygen also reacted with gases such as methane in the air, producing carbon dioxide and water, while it mopped up any free hydrogen produced by volcanoes, making yet more water. When all the things it could react with were gone, oxygen built up in the air, until by about a billion years ago the atmosphere began to resemble closely the mixture of gases we breathe today – 78 per cent nitrogen, 21 per cent oxygen and just small traces of everything else, including carbon dioxide.[1]

Animals take advantage of this free supply of oxygen by breathing it in and allowing it to combine with things like hydrogen and carbon in a kind of very slow burning process that provides the energy we use to live. So the evolution of animal life forms, breathing oxygen and using chemical energy in a profligate fashion (compared with plants), only occurred after the spread of oxygen in the atmosphere.

Respiration, as the oxygen-burning process is called, provided such an enormous advantage that, after one and a half billion years in which all life on Earth was the single-celled kind, a variety of multicellular animal life exploded on to the evolutionary scene with great rapidity, in the space of about a hundred million years, a little less than two billion years ago. The reason for their success is that the chemical energy available from respiration (slow burning) is quick and easy to obtain – but it is only available

[1] But remember that 60 atmospheres of carbon dioxide is locked up in the rocks today, so that in terms of what has been outgassed by volcanoes since the planet formed, the entire present-day atmosphere is only 1.6 per cent of what it might have been if the Earth had ended up like Venus.

because photosynthesis in plants has used that slower and more complicated process to take energy from sunlight and use it to make, among other things, free oxygen. Even plants today also use respiration, as well as photosynthesis, in their chemical processes; but in a neat example of the way evolution has to build on existing forms, instead of going back to the beginning and starting again, they still throw away the oxygen released during photosynthesis, then breathe oxygen in from the air during respiration.

The emergence of life out of the oceans and on to the land also owed its existence to the presence of oxygen in the atmosphere, but for quite another reason. Before there was oxygen in the air all the energy in sunlight could reach the surface of the Earth. A lot of this energy is in the form of ultraviolet radiation, and ultraviolet radiation is harmful to life – specifically, the photons of ultraviolet light have just the right amount of energy to split apart some of the chemical bonds in DNA molecules (which is why UV can cause skin cancer). But UV also reacts with oxygen, in a series of reactions which produce a form of oxygen with three atoms in each molecule, not two.[2] In essence, the UV breaks apart individual oxygen molecules, producing two free atoms which can each latch on to another oxygen molecule. The tri-atomic form of oxygen is called ozone, and as a result of these reactions it has built up in a layer of the atmosphere between about fifteen kilometres and fifty kilometres above sea level, the stratosphere (but don't be fooled by this great range of altitudes; the air is so thin in the stratosphere that if all the ozone were brought down to sea level, the pressure of the atmosphere would squash it into a layer only three millimetres thick). Ozone itself also absorbs UV radiation, and the concentration of ozone in the stratosphere stays roughly in balance as ozone is both created

[2] This is possible because of a quantum resonance, like the one which makes the carbonate ion stable, discussed in chapter four. If one oxygen atom gives an electron to another, it is left with a positive charge and an outer shell of five electrons, so it can make three covalent bonds, one with the oxygen atom that gained the extra electron and now has seven in its outer shell, and two with an ordinary oxygen atom. The resonance ensures that, in fact, each of the bonds in an oxygen molecule are like one and a half ordinary bonds. Without quantum resonance, there would be no ozone layer (among other things!) and we would not be here.

and destroyed by reactions involving sunlight – rather like a bath in which the taps are running but water is leaking away down the plughole, so that the level of water in the bath stays the same even though the individual molecules of water in the bath are constantly changing.

Life emerged on to land under the protective layer of ozone, and life itself maintains the protective layer of ozone as photosynthesis continues to put oxygen into the air, while the respiration of animals (and plants) and other oxidising processes (such as forest fires) remove oxygen from the air. At the end of the twentieth century, there is, rightly, considerable concern about the damage that gases manufactured by human industrial processes, and which have never existed naturally (the CFCs), are doing to the ozone layer. These gases are extremely inert, so they are not destroyed by any chemical or biological processes that occur near the ground. Eventually, they make their way into the stratosphere, where their molecules are broken up by the action of UV (another example of just how powerful and damaging UV photons can be) and release chlorine, which destroys ozone. The notorious hole in the ozone layer over Antarctica, and its counterpart over the Arctic, are produced directly as a result of the presence of CFCs in the atmosphere. Although the release of these harmful compounds has been slowed, there is so much potentially damaging material already in the air and working its way up into the ozone layer that it will take decades for the ozone layer to recover, even if no more CFCs are released.

The way in which living organisms interact with the atmosphere in a feedback process which maintains roughly the mixture of gases that we breathe today is fairly obvious, once it is pointed out. Slightly more surprisingly, though, this kind of feedback process also links life on Earth with the tectonic activity of our planet. Carbon dioxide from the air dissolves in rainwater, forming carbonic acid, which eats away at rocks that contain compounds of calcium, silicon and oxygen (calcium silicates). This chemical action releases calcium and hydrogencarbonate ions (a hydrogencarbonate ion is, logically enough, a carbonate ion with one hydrogen atom attached to it), which eventually get into the sea and are used by living organisms, such as plankton, to build their chalky shells, which are chiefly made of calcium carbonate. When these creatures die their shells fall to the sea floor, building up layers of sediment rich in carbonate. This process must have thinned the carbon dioxide atmosphere of the early Earth. But as geological activity carries the thin crust of the sea floor under the edges of the thicker continental crust, the carbonate is pushed under the continents and deeper into the Earth, where it gets hot and melts. At high temperatures and pressures carbon dioxide is released as new silicate rocks are formed, and the gas finds its way up to the surface and out into the atmosphere during volcanic eruptions.

These geological processes take a very long time to cycle carbon around. Once the system is running, though, the amount of carbon dioxide being released into the atmosphere by volcanoes in each millennium is roughly constant, and in the long run this is in balance with the amount of carbon dioxide being lost through the combination of geological and biological processes. But what happens if the temperature of the globe changes? If the temperature falls, less water evaporates from the oceans, so there is less rain, less weathering of the rocks, and less carbon dioxide taken out of the air. As the output from volcanoes continues, unabated, the concentration of carbon dioxide will increase, strengthening the greenhouse effect and warming the Earth, thereby increasing rainfall once again until equilibrium is restored. Similarly, if the world warms for some reason (such as an increase in the heat output of the Sun), there will be more rainfall and more weathering, which takes carbon dioxide out of the air and reduces the greenhouse effect.

This is an example of a negative feedback, a stabilising influence on the Earth's environment. Increasing the temperature of the Sun reduces the strength of the carbon dioxide greenhouse effect. Life also turns out to play a part in this geological stabilising process. The way the climate of our planet has evolved and the way life has evolved are so intimately connected that the process is sometimes described as the 'co-evolution' of climate and life.

More daringly, James Lovelock has suggested that the best way to understand these processes is to treat the entire planet, including both geological and biological activity, as a single living organism, which he calls Gaia. This is still a controversial idea, even though it has been around since the 1970s. But even without taking on board the idea that the Earth is literally alive, planetary scientists have found that Lovelock's approach, which takes account of the biological contribution in geological and climatic feedbacks of many kinds, is a fruitful way to get an understanding of how the environment of our planet is adjusted by a variety of natural checks and balances.

It now seems to be established beyond doubt that at least what you might call a 'weak' version of the Gaia idea holds – that life is not only affected by the physical environment, but also affects the physical environment, even on geological timescales, and often (indeed, usually) in such a way that conditions suitable for life are maintained. The various cycles involving carbon dioxide are among the most important aspects of this relationship between living things and the physical environment, and by digging fossil fuel out of the ground and burning it, as well as by cutting down tropical forests which take carbon dioxide out of the air during photosynthesis, we are upsetting the natural balance of a key process involved in regulating the temperature of the Earth.[3] Before we look in a little more detail at this human (or 'anthropogenic') greenhouse effect, though, we ought to take a quick look at what the climate was doing before human intervention.

We have already mentioned how the climate of the Earth has been affected by the long, slow processes associated with continental drift, with ice ages occurring when land drifts across one or both of the poles. But now we want to concentrate on the way the Earth is today and has been for the past several million years, during the time when our own species, *Homo sapiens*, has emerged on to the evolutionary stage.

The driving force of the whole weather machine is, of course, the Sun; but the main input of energy from the Sun occurs not at the top of the atmosphere, but at the bottom. Most of the

[3] Of course, since we are living things that have evolved on Earth, in a sense anything we do is also 'natural', but we trust our meaning is clear.

energy coming in from the Sun passes through the atmosphere unimpeded, and warms the surface of the Earth. It is the warm surface of the Earth (land or sea) that warms the *bottom* of the atmosphere in its turn. So the warmest layer of air is the one at ground or sea level, and, to start with, as you go up through the atmosphere the temperature drops. The average temperature at sea level is about 15°C, but this falls off so rapidly that at the top of the stratosphere, roughly fifteen kilometres above sea level, it is down to a chilly –60°C.

In the stratosphere (the ozone layer) the temperature increases once again, because energy in the form of ultraviolet radiation from the Sun is being absorbed. So at the top of the stratosphere, some fifty kilometres above sea level, the temperature of the air is back to the freezing point of water, 0°C. There is another cooling layer above the stratosphere. This is known as the mesosphere, which extends to an altitude of more than eighty kilometres. At higher altitudes still, the tenuous gases around the Earth are affected by energetic particles emitted by the Sun, and cosmic rays from the depths of space, and scarcely correspond to what we would think of as air. What matters in terms of driving the weather machine is the troposphere, the layer in which weather occurs, and the stratosphere, which acts like a lid on the troposphere.

The stratosphere acts like a lid because it is a layer in which temperature increases with altitude. The old adage that 'hot air rises' is true only if the air above the hot air is cooler than the rising air. When heat from the Sun warms the surface of the Earth, the hot air at the surface rises through the troposphere, expanding and cooling as it does so. It cannot rise into the stratosphere, where the temperature is higher than at the top of the troposphere.[4]

This solar heating is greatest in the tropics, where the Sun is almost directly overhead at noon. At higher latitudes, the Sun is lower on the horizon, so the heating effect is weaker, just as its

[4] That is, large amounts of air moving in bulk cannot penetrate the stratosphere; individual molecules in the air can cross the barrier by diffusion, which is how CFCs get up there and do their damage.

heating effect is weaker early in the morning or late in the evening than it is at noon at any latitude. The reason for this is that when the Sun's rays strike the surface of the Earth at a shallower angle they are spread out over a wider area – you can see what is going on by shining a torch at a piece of cardboard. If the cardboard is held perpendicular to the beam of the torch, the light makes a small but bright round blob on the cardboard. But as you tilt the cardboard away from the beam, the light spreads out to make an elongated but fainter blob. The same amount of light covers a bigger area, so each part of the elongated blob is fainter, even though the total amount of light is the same.

The hot air rising in the tropics is also moist, because it contains water vapour evaporated from the sea. As the air rises and cools, this moisture falls as rain, which is why the tropics are wet. The air that has risen from the surface turns over high in the troposphere and pushes out on either side of the equator. Once it has cooled, it falls back down to the ground. The descending air gets hotter again, because it is being squashed by the pressure of the air following behind it – the same effect is at work in a bicycle pump, making the air in the pump hot as you force it into the tyre. The hot air absorbs moisture, so where it reaches the ground there are bands of desert on either side of the equator – including the Sahara desert. Some of this dry air then flows back to the equator along the surface of the Earth, producing the trade winds and picking up moisture as it does so; some is involved in more convective cycles, which together carry heat away from the equator and out towards the poles, producing the prevailing winds at higher latitudes, including the westerlies which come in from the Atlantic and dominate the climate of Britain.

Climate is a kind of average weather, produced by the combined effects of all these processes over many years. But climate is always changing, on a variety of timescales, because the various inputs to the weather are always changing. They even change during the course of the year, although we are so used to this that we do not think of it as a climatic change.

The Earth is essentially a ball of rock, covered by a thin smear of atmosphere and ocean, which turns on its axis once every twenty-four hours and orbits around the Sun once every year.

But the axis on which it rotates every twenty-four hours (an imaginary line joining the North Pole to the South Pole) is tilted relative to the plane in which our planet orbits around the Sun. This tilt stays the same over the course of a year, always pointing in the same direction; but as we move around the Sun sometimes we are on one side of the Sun and sometimes (six months later or earlier) we are on the opposite side of the Sun. So sometimes the North Pole is tilted towards the Sun, and sometimes it is tilted away from the Sun, with the South Pole always being affected in the opposite sense. Exactly the same process that makes higher latitudes cooler than lower latitudes – the spreading out of incoming solar energy across a larger area when the Sun is lower in the sky – makes the hemisphere that is tilted away from the Sun cooler than the hemisphere that is tilted towards the Sun. This is why we have the cycle of the seasons, and why the Sun rises higher in the sky in summer than it does in winter.

Although the tilt of the Earth can be regarded as constant over an interval as short as a year, the spinning planet is actually like a wobbling gyroscope, and anyone who has watched a child's spinning top in action will know how a gyroscope can wobble as it spins. The first wobble of the Earth follows a circular path, largely because of the gravitational influence of the Sun and Moon, tugging on it as it spins; but the wobbling is a very slow process compared with the cycle of the seasons. The wobble makes an imaginary line, pointing straight upward from the North Pole, trace out a circle on the sky, following a cycle which varies in length but takes about twenty thousand years to complete. This shifting of the direction in which the pole points (called precession) causes an apparent shift of the stars across the sky as seen from Earth. One effect of the wobble is that the pattern of the seasons changes slowly as the millennia pass. Another is that ten thousand years ago the present-day Pole star would not have been a good guide to the direction of north.

At the same time that this precession is going on, the angle of the Earth's tilt changes, gently nodding up and down as the millennia pass. The way this nodding occurs can be calculated very easily from Newton's laws. It varies between being 21.8 degrees out of the vertical and 24.4 degrees out of the vertical, over a cycle roughly 41 thousand years long. Today the tilt is 23.4 degrees, and is decreasing. Other things being equal, this means that the differences between summer and winter are less extreme today than they were a few thousand years ago. Summers are a little cooler, and winters are a little warmer, than they used to be.

A third astronomical influence on climate is caused by the interacting web of gravitational forces between the Sun and all the planets of the Solar System, which stretches the Earth's orbit slightly, from more circular to more elliptical, and then squeezes it back towards being circular, over a period a little over a hundred thousand years long. When the orbit is more or less circular, the planet receives the same amount of heat from the Sun each day. When the orbit is more elliptical, during one part of the year the Earth is closer to the Sun, and gets more heat, while at other times it is further from the Sun, and gets less heat. But the important point about all of these astronomical influences on climate is that they do not change the total amount of heat received by the whole Earth over a whole year. All they can do is rearrange the distribution of heat between the seasons.

It happens that these effects, although small, have a strong influence on climate today because of the present geography of the globe. There is ice over both poles, which is unusual in itself. The ice over the South Pole is very much the kind of thing that has happened earlier in the history of the Earth, with a continent – Antarctica – sitting squarely over the pole and blocking the flow of warm water to high southern latitudes. But the ice cover in the north is the result of a much rarer geographical arrangement (possibly unique in the history of our planet), and is much more interesting.

The Arctic is also ice-covered because warm water cannot easily penetrate to high latitudes. But this time it is because the Arctic Ocean is almost surrounded by land. The pole itself is covered by water, and the ice floats on the surface of the Arctic Ocean. It turns out that this makes the northern hemisphere very sensitive to the astronomical rhythms associated with the wobbling of the spinning top Earth.

This idea of an astronomical influence on climate was first put forward by the Scot James Croll in the 1860s, and was refined by the Yugoslav astronomer Milutin Milankovitch in the first half of the twentieth century – it is often called the Milankovitch Model. It was only proved to be a good model in the 1970s, though, when a combination of electronic computers, to calculate precisely how the astronomical influences have changed in the past hundreds of thousands of years, and studies of cores drilled from the deep sea bed, which show how climate has changed over the same interval of time, showed an almost perfect match. The Milankovitch Model, like all good models, agrees with experiment.

What the combination of observations and theory tells us is that, given the present-day geography of the globe, ice ages occur when summers in the northern hemisphere are particularly cool. This looks odd at first sight, because cool summers, as we have seen, go hand in hand with relatively warm winters. You might guess that in order to encourage ice to spread out of the Arctic region and across the land which surrounds the Arctic Ocean, you would need cold winters to make lots of snow. But remember that the average amount of heat received by the Earth over a year is the same all the time. So cold winters go hand in hand with hot summers, and hot summers are good at melting snow. The key point is that the Earth today is always cold enough for snow to fall in winter on the land around the frozen Arctic Ocean. In order for the ice to spread, what matters is for the summer temperature to be so cool that some of this snow fails to melt, and builds up from year to year, eventually building new ice sheets.

But, for once, 'eventually' may not be very long. If even a thin layer of snow lasts through the summer, it has a big impact on the local climate because it reflects incoming solar heat that would otherwise have warmed the ground below. At the start of the next winter the ground is cooler than it would otherwise be, and it is easier for the next fall of snow to lie there without melting. So the natural balance quickly changes into a full ice age when the conditions provided by the Milankovitch Model are right; the switch almost certainly takes less than a millennium, and perhaps only a couple of centuries. But warming the planet out of an ice age is much more difficult, because of the way ice and snowfields reflect away incoming heat. It takes several thousand years to make the switch from full ice-age conditions to a full interglacial.

In fact, the way we have told the story, although natural from the point of view of people living in a relatively warm interval, is the wrong way round. With the present-day geography of the Earth, the natural state of the planet is to be in the grip of a full ice age. It is only when all the Milankovitch cycles work together to bring the maximum amount of summer heating that enough of the ice is melted in the northern hemisphere to bring a brief respite. And, as we have mentioned, at present summers are getting cooler.

With three different cycles operating together to change the climate in this way, and each of them being subject to slight variations in their own right, the pattern does not repeat exactly. Roughly speaking, though, the pattern of climate change over the past five million years or so has been one in which full ice ages, each a bit more than a hundred thousand years long, are separated by shorter, relatively warm intervals, each about ten or twenty thousand years long. These warm intervals are called interglacials, and we are living in an interglacial that began about fifteen thousand years ago. It is still colder than the Earth has been throughout most of its history – ice at either pole is rare on a geological timescale. All of human civilisation has developed during this single interglacial – and it may be *because* of this ice age/interglacial cycle that we are here at all.

It wasn't so much the cold as, almost certainly, the drought associated with a full ice age that drove our ancestors out of the East African forests and set them on the path to being human. Drought goes hand in hand with the spread of ice, both because water is locked up in ice and because when the world is cold less water evaporates from the oceans, so less is available to fall as rain.

At the time the repeating rhythm of ice ages and interglacials became established, our ancestors were leading quite comfortable and successful lives as inhabitants of the open forests and grasslands of East Africa, as testified by a wealth of fossil remains. But then the Earth was plunged into the succession of ice ages and interglacials that has operated ever since, and which seems to have been unique in the long history of our planet. Every time an ice age set in, the forests were struck by drought, and went into decline, while the plains became more arid. These are just the sort of harsh conditions which give a turn to the evolutionary screw, with less adaptable species suffering most, while those that can cope with the changing conditions thrive. During the interglacials, though, wetter conditions returned as the world warmed, so that the trees and grass (and plant life in general) flourished, and those individuals and species that had survived the drought had an opportunity to spread out, enjoying a population boom.

Every time there was drought, only the most adaptable survived; every time the rains returned, the descendants of those adaptable individuals flourished. There are two ways in which the evolutionary adaptation could proceed under these conditions, starting out from, for want of a better term, an 'apeman'. One successful line would be to become ever more closely adapted to the remaining forests, retreating into the heartland of the woods whenever the drought struck. Such an evolutionary development would lead to the modern chimpanzee and gorilla. The other successful strategy would be to become increasingly well adapted to the dry plains that would spread as the forest shrank. As has happened so often in the evolutionary past, it would be the *less* successful creatures that would be forced to the fringes – in this case, the less successful apes that would be forced out of the shrinking forest and left to fend as best they could on the plains. If the drought had lasted for a million years, maybe the reluctant plains dwellers would have died out entirely; but after each hundred thousand years or so of drought, the survivors would have had a chance to regroup and increase in numbers when the rains returned, almost like a regiment being pulled out of the front line for rest and recreation before being returned to the battle. It is clear that under these conditions, coping with changing environmental surroundings, intelligence would have a distinct evolutionary advantage.

It isn't just the fossil record that bears out this admittedly speculative, but plausible, scenario. In round terms, human DNA is more than 98 per cent the same as the DNA of chimps and gorillas (and any two human beings share a little more than 99.8 per cent of their DNA). We are only 1 per cent human, and roughly 99 per cent ape. Molecular biologists can measure the rate at which changes accumulate in DNA over the generations, by comparing DNA from many living species with fossil evidence of when different lines split from a common ancestor.

The amount of difference between ourselves and the other African apes points to a three-way splitting of the evolutionary line from a common ancestor that was around about five million years ago. The split occurred exactly when the distinctive pattern of climatic changes that we think of as normal set in. This split is so recent that it seems likely that the common ancestor of ourselves, the chimpanzees and the gorillas was more 'man-like' than 'ape-like' – in particular, it had already developed an upright stance. It was the pressure of repeated drought and the need to adapt ever more closely to a forest life that made the chimp and gorilla what they are today, so that, turning the usual vernacular expression on its head, rather than describing humankind as being descended from the apes, it would make more sense to say that the apes are descended from man – or, at least, from proto-man.[5]

We owe our existence to a kind of ratcheting effect, resulting from these natural climate rhythms. The natural trend would be for these Milankovitch rhythms to plunge the world back into a full ice age within the next few thousand years. But human activities are now a significant part of the weather machine, and it seems likely that the next ice age may be postponed indefinitely as a result, either by accident or design.

The concentration of carbon dioxide, the principal greenhouse

[5] This idea was first proposed by John Gribbin and Jeremy Cherfas in the early 1980s, and elaborated on in their book, *The Monkey Puzzle*. It was regarded as a wild speculation at the time; however, improved molecular studies involving DNA from different species made the idea respectable in the mid-1990s, when it was taken up by Simon Easteal and Genevieve Herbert, of the Australian National University. It is now well established that, in their words, 'the common ancestor of humans and chimpanzees was bipedal and that the trait has been lost in chimpanzees rather than gained in humans'. (*Journal of Molecular Evolution*, February 1997)

gas, in the atmosphere is only about 0.03 per cent, or 300 parts per million (ppm). Since the industrial revolution of the nineteenth century, however, it has increased from 280 ppm to 350 ppm, entirely as a result of human activities – burning fossil fuel, and destroying forests. This is a relative increase of 25 per cent, and is already changing the world's weather. The world has warmed overall by about a degree (Celsius) since the 1880s, and carbon dioxide is continuing to pour into the atmosphere, at an increasing rate. Just how far the warming will go depends on how much more carbon dioxide gets into the air, and how quickly; but conservative estimates suggest a warming of a further degree over the next twenty years or so, after which projections are more guesswork than science. Even this seemingly modest warming, though, would take temperatures above anything that has been experienced during the present interglacial, and therefore (since, obviously, temperatures were lower in the latest ice age) higher than anything that has been experienced on Earth for well over a hundred thousand years.

The weather of the twenty-first century will be unlike anything that has occurred on Earth since civilisation began, and during the first half of that century the world will be warming some fifty times faster than it did at the end of the latest ice age, when the change from full ice age to full interglacial conditions involved a warming of about 8°C over about five thousand years. There is just no way to tell what effect so dramatic a change in the physical environment will have on the living environment. So, taking the coward's way out, we will make no attempt to predict the future climate of our planet, and will instead take our next step out into the Universe at large, where we can tackle the much simpler problems of explaining what it is that makes the Sun and stars shine, and where the Sun got its family of planets from.

CHAPTER NINE

THE SUN AND ITS FAMILY

The Sun is a star. It is a fairly ordinary star, neither particularly large nor particularly small, neither particularly bright nor particularly dim, roughly halfway through its life cycle. The only reason that it looks any different from the other stars in the sky is that we are so close to it – the Earth orbits round the Sun at a distance of only 150 million kilometres, taking one year to complete the journey. Most astronomy books (and most schoolteachers) will tell you that there are nine planets, including the Earth, in the Sun's family. This is, however, misleading, because the most distant of these nine objects from the Sun, Pluto, is clearly a different kind of object from the other eight planets, and is better described as a rather large piece of cosmic debris, more like the comets and asteroids that litter the Solar System than a true planet.

One of the most obvious features of the peculiarity of Pluto compared with the other planets is that although on average it is indeed the most distant of the nine recognised planets from the Sun, it has a more elliptical orbit than any of the other planets so that some of the time (including between 1979 and 1999) it is actually closer to the Sun than Neptune, the most distant of the other planets from the Sun. No other planet crosses the orbit of any of its neighbours.

Astronomers measure distances across the Solar System in terms of the average distance of the Earth from the Sun, which is defined as one astronomical unit (1 AU); Pluto's average distance from the Sun is just under 40 AU, but the actual distance varies between 30 AU and 50 AU in different parts of its orbit. So at its most distant from the Sun, Pluto is fifty times further out than we are. It is only 2,320 km in diameter (two-thirds of the size of our Moon) and has a mass only 0.3 per cent of the mass of the Earth. Pluto is accompanied by a moon called Charon, which is 1,300 km across (more than half the size of Pluto) and orbits Pluto at a distance of only 19,400 km. Both objects are made largely of water ice and frozen methane, with an average density less than twice the density of water. Their surfaces are at a chilly 50 Kelvin (below *minus* 220 degrees Celsius), and they take 248 years to orbit around the Sun once.

With Pluto out of the way, we can concentrate on the eight real planets which form the most obvious members of the Sun's family, as well as taking a closer look at the rest of the cosmic debris in the Solar System. The family divides neatly into two groups of four – four rocky, small planets orbiting in the inner part of the Solar System, and four large, gaseous planets orbiting in the outer part of the Solar System, separated by a band of rubble (between the orbits of Mars and Jupiter) known as the asteroid belt.

All bright stars shine, like the Sun, because they generate heat in their interiors, through the process of nuclear fusion. The planets, though, are only visible because they reflect light from the Sun. As a result, they are much fainter objects, and this is why until recently there has been no direct evidence for the existence of planets around other stars – astronomers were sure they must be there, but the planets were too dim to be seen. In the 1990s, though, direct evidence was found for the existence of planets orbiting other stars, from extremely accurate measurements of the way those stars seem to wobble about on the sky. The wobbling is interpreted as the effect of the gravity of a large planet tugging on the parent star as it orbits around the star, pulling it first one way and then the other. The technique only reveals the presence of larger planets (like Jupiter, the largest planet in our Solar System), but it is a reasonable extrapolation to infer that if planets like Jupiter are seen orbiting around stars like the Sun, then planets like the Earth are probably in orbit around at least some of those stars.

Another development in the 1990s has been the direct detection of clouds of dusty material in a fat disc around some young stars. Even before such discs were identified (and photographed directly using the Hubble Space Telescope and other instruments), astronomers had what seemed to be a good model of the way in which the planets of the Solar System could have formed from a dusty disc surrounding the young Sun. The existence of exactly the kind of discs required by that model photographed in orbit around stars that are exactly like the young Sun (according to astrophysical models, which we discuss in more detail in the next chapter) leaves little room to doubt that we really do understand how the Sun and its family formed. But this is not quite the same as testing models by experiments carried out in the laboratory – we cannot test our model of how the Solar System formed by making another Solar System.

This highlights an important difference between astronomy and most of the science we have described so far in this book, and it means that to a degree astronomical models are always less satisfactory than the best models we have for processes that go on here on Earth. But that degree can be very small in some cases, and, as we hope we shall make clear, it certainly does not mean that all astronomical models are no more than wild speculations (*some* are, but we won't discuss them here!). The best of those models are tested by comparison with the way things behave in the real Universe, by computer simulations, and in many cases (including, for example, the way the stars generate heat) by using data from experiments in laboratories on Earth which mimic at least some of the key processes involved in the astrophysical phenomena of interest.

The best model we have of the way in which the Sun and its family of planets (and other bits of debris) formed relates these events to the structure of the whole galaxy of stars (the Milky Way Galaxy) in which we live. The realm of the stars is discussed in more detail in chapter ten, but the relevant features are that the Sun is one of a few hundred billion stars which together form a flattened, disc-shaped system roughly a hundred thousand light years across and a couple of thousand light years thick.[1]

[1] A light year is, literally, the distance light can travel in one year. It is a measure of distance,

The Solar System is about two-thirds of the way out from the centre to the edge of this disc and, like the rest of the disc population, orbits around the centre of the disc. Travelling at a speed of 250 kilometres per second, it takes us about 225 million years to complete one orbit – an interval sometimes known as the 'cosmic year'. Like many other disc galaxies, our own Milky Way is marked by two distinctive features known as spiral arms, which coil outwards from the centre – they are revealed by the radio emission from clouds of hydrogen gas which they contain.[2] The spiral patterns are pressure waves, and everything orbiting around the Galaxy gets squeezed when it passes through them. The equivalent arms in other galaxies show up brightly because they contain many hot, young stars, and they contain hot, young stars because clouds of gas and dust orbiting around the galaxy get squeezed by the pressure wave when they pass through a spiral arm. In addition, the more massive stars that line the spiral arms run through their life cycles quickly and explode, sending out blasts which put an extra squeeze on any nearby clouds of gas and dust.

The stars of the disc form in groups, with many stars being born together out of a single large, collapsing cloud of gas and dust. At first, they form what is known as an open cluster of stars, and there are more than seven hundred of these open clusters within about eight thousand light years of the Sun. But these clusters are not held together strongly enough by gravity to maintain their identity, and the individual stars get spread out and move away from one another as they orbit around the galaxy, so that after a few hundred million years there is no longer any way to tell which stars were born together, out of the same cloud.

Our Sun formed in this way about five billion years ago, and has been orbiting the Galaxy ever since, completing twenty or so orbits in its lifetime to date. The cloud of stuff from which the

not of time, equivalent to 9.46 million million km. For comparison, light covers the 150 million km from the Sun to the Earth in 499 seconds. So the distance from Earth to the Sun is 499 light seconds, or 8.3 light minutes.

[2] The spiral pattern of a galaxy like our own Milky Way is superficially similar to the spiral pattern made by cream as it is stirred into coffee.

Sun and its family (along with the other stars of a now widely dispersed open cluster) formed was almost entirely made of hydrogen (about 75 per cent) and helium (about 25 per cent), gas left over from the Big Bang in which the Universe was born (*see* chapter eleven). But it was laced with a smattering (about 1 per cent) of heavier elements, all of them manufactured inside stars (as we describe in the next chapter) and scattered through space when those stars died. When the fragment of interstellar cloud that was to become the Solar System began to shrink, collapsing under its own weight, it got hot in its heart, because gravitational energy was released as the cloud got smaller – if two things are attracted to one another by gravity, obviously you have to put energy in to move them further apart; when they move close together, the same amount of energy is released, and this applies to every molecule of gas in the shrinking cloud. Eventually, it got so hot inside (about 15 million degrees Celsius) that hydrogen nuclei began to be converted into helium nuclei, releasing energy as they did so.

In a multi-step process four protons (hydrogen nuclei) are combined to make one alpha particle (a helium nucleus). The total mass of the alpha particle is just 0.7 per cent less than the mass of four protons added together, and this much mass is converted into pure energy every time the process is completed. In order to stabilise the Sun and stop it collapsing any further today, five million tonnes of mass are converted into pure energy in this way every second (roughly the equivalent of converting a million elephants into pure energy every second). Even after five billion years of producing energy at this prodigious rate the Sun has so far used up only about 4 per cent of its initial supply of hydrogen, with just 0.7 per cent of that 4 per cent actually having been turned into radiation and lost into space. The mass equivalent of all of the energy radiated by the Sun in its entire lifetime to date is about a hundred times the mass of the Earth.

In about another five billion years or so, the Sun will begin to run into problems, because it will have used up all the hydrogen in its core. There will still be hydrogen in plenty in the outer layers of the star, but the hot core itself will be composed almost entirely of helium, the 'ash' of this lifetime of nuclear burning. At that time in its life, the Sun's core will shrink and get hotter still, and this will allow other nuclear fusion reactions to take place, converting helium nuclei into carbon nuclei at a temperature of about 100 million degrees Celsius. The extra heat generated in the core at this stage of its life will make the outer layers of the Sun swell up, turning it into a kind of star known as a red giant, and engulfing the innermost planet, Mercury. Eventually, after a further billion years or so, when its supply of helium is also exhausted, the Sun will cease to generate energy in its interior, and will fade away to become a cooling cinder, no bigger than the Earth itself, called a white dwarf. More massive stars run through their life cycles more quickly, because they have to burn fuel more vigorously to hold themselves up against the inward tug of gravity; they also do more interesting things at the end of their lives, as we describe in chapter ten.

From the point of view of the Sun's family of planets (and ourselves), though, what matters now is that for a full ten billion years a star like the Sun generates energy at a more or less steady rate, while the planets wheel around it and develop in their own particular ways, including (on at least one of those planets) the development of life. But why are the planets so different from one another?

The nature of the individual planets of the Solar System, as it condensed out of a collapsing cloud of gas and dust, was determined first by rotation and then by the heat being radiated from the Sun itself, once it had formed. Any cloud of material in space is bound to be rotating – the chances of it being poised in a stationary state are negligible. As the cloud began to collapse inward it would have rotated faster, in just the same way that a spinning ice skater can increase their speed of rotation by pulling their arms inward. This is due to the conservation of what is called angular momentum. The angular momentum of a mass moving round in a circle depends on the amount of the mass, its distance from the centre of the circle, and the speed at which it is moving. So if the same mass gets closer to the centre it has to move faster to conserve angular momentum. Most of the mass in the cloud that condensed to form the Solar System settled into a ball at the centre – the Sun.[3] But this was only possible because the angular momentum of the cloud largely got transferred into a disc of material left behind around the central ball of gas. By spinning faster, and also being further out from the centre, this disc was able to store most of the original angular momentum from the material that became the Sun. The Sun got most of the mass, but the disc got most of the angular momentum.

The planets and moons of the Solar System formed in this swirling disc, and preserve its original angular momentum. All of the planets orbit around the Sun in the same direction, and almost all of the moons orbit around those planets in the same direction as well. Even the spin of the planets, with the exception of Venus and Uranus (which seem to have been jolted by major cosmic impacts), is in the same direction, storing a little bit more of the original angular momentum, and the Sun itself, rotating once every 25.3 days, also spins in the same direction. This is clear evidence that the Sun and planets formed together from a single rotating gas cloud, and that the Sun did not, for example, pick up the planets one by one as it orbited around the Galaxy. If that had been the case, their orbits would be inclined at random, not in a single disc, and the direction of the planets in those orbits would also be random.

As well as the direct evidence we now have for discs around young stars, the two largest planets in the Solar System, Jupiter and Saturn, are themselves like miniature Solar Systems, escorted by families of moons which orbit around them in the same way that the planets orbit around the Sun. These giant planets clearly formed in the same way as the Sun but on a smaller scale, growing discs of material around themselves as they contracted, and forming both moons and rings out of the debris in which angular momentum is stored.

The planet-forming process would have started even before the central ball of gas that was to become the Sun had got hot enough to ignite nuclear fusion. Tiny pieces of dust which were in the

[3] That is, most of the mass that collapsed; some of the original mass got blown away into space by the heat of the cooling cloud, or pushed away by magnetic fields, helping to carry angular momentum away.

original cloud would have stuck together to make little fluffy grains a few millimetres across, and these grains would have collided with one another and stuck together to make still larger grains. In the early stages of this process, the grains would have been immersed in gas, constantly bombarded by molecules of gas in the collapsing cloud, and these collisions would have ensured that angular momentum was shared out, with the material settling into a disc around the proto-Sun. This concentration of material into a disc would have made collisions between the particles more likely, so, even though the remaining gas was blown out of the Solar System as the Sun began to heat up (probably as much gas as remains in the Sun today was lost in this way), the supergrains that had been built up in the cloud would still be able to interact with one another.

The process of accretion continued in much the same way until objects the size of asteroids today had been built up – rocky lumps maybe a kilometre or so across. By then, though, gravity had begun to be important, tugging these lumps of rock together into swarms where they could bump into one another, sticking together in larger lumps. The largest lumps, with the strongest gravitational pull, would attract more material onto themselves, increasing their mass and gravitational pull still further and growing to become the planets. At this stage the heat generated by the impacts of successive waves of rock pounding the protoplanet would have melted it right through, allowing iron and other metals present to settle in the core and, as the planet cooled, producing the kind of layered structure that we see in the Earth today.

On this picture it is easy to explain why there are two kinds of planet and a lot of cosmic debris in the Solar System today. Close to the Sun, the heat from the young star would have driven away light gases and material that could easily be vaporised. The grains that could survive in this heat would be rich in material which does not easily vaporise, such as iron and silicates. These formed the building blocks of the inner planets, which are small and rocky, and have only modest cloaks of atmosphere.

Further out from the young Sun, the grains from which the planets formed would retain a coating of water ice, frozen methane, and solid ammonia (all substances which we know exist in interstellar clouds, from spectroscopic studies). In addition, the very light gases, hydrogen and helium, blown away from the inner Solar System, would be available to be attracted by the gravity of any planet that formed in these colder regions. So the outer planets, although they may have grown initially because of the gravitational attraction of rocky lumps that formed in the same way as the inner planets, are almost entirely made of gas, and have only relatively small rocky cores.

One important feature of this planet-building process is that large amounts of cosmic rubble must have been left over after the planets had formed. A lot of this rubble is around today in the asteroid belt, between the orbits of Mars and Jupiter. It could not clump together to form a planet there, because of the disturbing influence of the gravitational pull of Jupiter itself. Further out from the Sun, it was – and is – cold enough for icy lumps of debris to remain, in the form of comets.

When the Solar System was young, in the billion years or so after the planets formed, cosmic debris was everywhere, and impacts frequently scarred the surfaces of the young planets. The dramatic cratering of the Moon is a reminder of this era of cosmic collisions, but Mercury, Venus, Mars and even the Earth itself also bear similar scars. Even though the major battering of the inner planets ended some four billion years ago, stray pieces of cosmic debris still collide with planets from time to time, as was dramatically demonstrated when the pieces of comet Shoemaker–Levy 9 hit Jupiter in 1994. As we have mentioned, there is little doubt that similar impacts on Earth have affected the course of evolution, and that one impact in particular contributed to the death of the dinosaurs, 65 million years ago. The Solar System is far from being completely quiet even today; but it has settled down into a very stable state, with the Sun and each member of its family having their own distinctive individual characteristics.

The Sun is by far the dominant member of the Solar System. It contains 99.86 per cent of the mass of the Solar System, and holds everything else – planets, comets, asteroids, odd bits of gas and other debris – in orbit around itself, held in the Sun's gravitational grip, as described by Newton. Just one planet, Jupiter, contains two-thirds of the remaining mass, which, strictly speaking, puts even the Earth down in the category of 'other bits and pieces'. But it is natural, however, from our human perspective, to describe even the Sun, as far as possible, by comparison with our home planet.

In round numbers, the mass of the Sun is 330,000 times the mass of the Earth, and it has a diameter of 1.4 million km, 109 times the diameter of the Earth. This means that 109 Earths could sit side by side along a single diameter through the Sun;[4] but because the volume of a sphere is proportional to the cube of its radius (or to the cube of its diameter), the volume of the Sun is 109 cubed times the volume of the Earth – rather more than a million times the volume of our planet. Averaging over the whole of the Sun, its density is only one-third of the average density of the Earth – so the Sun is about 1.4 times the density of water. But in the heart of the Sun, where nuclear reactions are generating energy, the density is 12 times the density of solid lead, and the temperature is about 15 million °C.

We know about the conditions deep inside the Sun through a combination of studies. First, the astrophysicists can work out (using very simple, basic physics) how hot the Sun must be in its interior in order to radiate as much energy as we see, and to hold itself up against gravity. Complementing this approach, experiments using particle accelerators here on Earth, combined with quantum theory, tell us how the energy in the Sun's heart is generated – and an absolutely crucial ingredient in all this is the way that quantum effects allow protons to fuse together even at a temperature of 'only' 15 million degrees. The stellar models (more about them in the next chapter), combined with particle physics, specify only a narrow range of possibilities for properties such as density and temperature at different depths inside the Sun. Best of all, in recent decades astronomers have been able to monitor tiny ripples occurring in the surface of the Sun, the solar equivalent of earthquakes. Using solar seismology, they have

[4] Coincidentally, it would take 107 Suns, side by side, to stretch across the distance from the Sun to the Earth – across the radius of the Earth's orbit, not its diameter.

probed the interior of the Sun in much the same way that geophysicists use earthquake waves to probe the interior of the Earth, and they find that the internal structure really does match the stellar models. So what we are telling you here is all good science, tested by experiment.

Under the extreme conditions at the heart of the Sun, electrons are stripped from their atoms to leave bare nuclei of hydrogen and helium (protons and alpha particles). Because nuclei are so much smaller than atoms, the central core of the Sun acts like a perfect gas, with nuclei bouncing off one another in high-energy collisions. This central region of the Sun, only 1.5 per cent of its volume, contains half of its mass.

The energy produced in the core is largely in the form of high-energy photons, initially gamma rays, which, under those conditions of extreme density, can travel only a short distance before meeting and interacting with a charged particle (electron, proton, or alpha particle). Although these interactions gradually degrade the gamma rays into slightly less energetic X-rays, each photon bounces around inside the Sun, ricocheting from one charged particle to the next like a ball in some crazy pinball machine. Even though each photon travels at the speed of light, it follows such a tortuous, zig-zag path that it takes, on average, ten million years for it to work its way out to the surface layers. Travelling straight outward at the speed of light, the journey from the centre of the Sun to the surface would take just 2.5 seconds; but light actually travels ten million light years of zig-zagging to complete the 2.5 light second journey.

One result of this is that the overall state of the Sun today is actually a smoothed-out result of everything that has been going on in its interior for the past ten million years or so. Looking at light from the surface of the Sun tells us more about what was happening in its core ten million years ago than about what was happening in its core yesterday. But the sound waves that make ripples on the surface of the Sun travel right through it in minutes, so helioseismology tells us about the interior structure of the Sun today, which makes it doubly valuable as a test of the accuracy of the models.

The zone of radiation extends out to a million kilometres from the centre of the Sun, about 85 per cent of the way to the surface. There, the temperature has dropped to 500,000°C and the density is only 1 per cent of the density of water. Some nuclei are able to cling on to electrons under these conditions, and the photons themselves have been degraded to longer wavelengths and lower energies as a result of their repeated collisions with charged particles on their ten-million-year-long journey. The overall result is that the partially ionised gas in this region can absorb energy from the radiation. The hot material produced in this way rises by convection, carrying energy outward over the last 15 per cent (about 150,000 km, half the distance from the Earth to the Moon) of its journey from the centre of the Sun to its surface. The visible bright surface of the Sun, which has a temperature of about 5,500°C, is where the atoms release energy in the form of photons of light, which then take only 8.3 minutes to cross the further 150 million km to Earth.

All the light we see from the Sun comes from a layer only 500 km deep, the top 0.1 per cent of the Sun. But the Sun's influence extends further out into space, through a kind of solar atmosphere, known as the chromosphere, which blends into a region known as the corona, which extends for millions of kilometres into space and produces a stream of tenuous material blowing outward from the Sun – the solar wind.

The closest planet to the Sun is Mercury, which orbits at a distance of 0.39 astronomical units and takes 87.97 of our days to complete one journey around the Sun. Because the planet rotates on its axis once every 58.64 of our days, three days on Mercury last for two of the planet's years. Although Mercury is visible to the naked eye (shining, of course, by reflected sunlight) and was one of the planets known to the ancients, it is very difficult to see in the glare of the Sun, and most of the information we have about the surface of Mercury comes from the spaceprobe Mariner 10, which made three passes by the planet in 1974 and 1975. Mariner 10 sent back pictures showing a heavily cratered surface very reminiscent of the surface of the Moon. This came as a complete surprise to astronomers, but is now incorporated into the standard model of the way in which the planets formed, with a heavy bombardment of asteroids continuing for hundreds of million of years after the planets achieved more or less their present sizes.

There is essentially no atmosphere on Mercury; temperatures there range from 190°C in the full glare of the Sun to –180°C on the night side. With a diameter of 4,880 km, Mercury is intermediate in size between the Earth and the Moon, and has a mass about 5 per cent of the mass of the Earth.

As we mentioned in the previous chapter, Venus, the second planet out from the Sun, is very nearly the physical twin of the Earth. It has 82 per cent of the mass of the Earth, and a diameter at the equator of 12,104 km, compared with the Earth's 12,756 km. Because the surface of Venus is completely covered by clouds, even the best telescopes on Earth cannot reveal any surface features, and the superficial similarity of Venus and Earth encouraged science-fiction writers (and even some scientists) to speculate that those clouds might hide a steaming jungle rich in life. But, as we have seen, the runaway greenhouse effect has actually made Venus a searing desert, with surface temperatures above 500°C, atmospheric pressure up to 90 times the pressure at the surface of the Earth, and highly acid rain falling from those clouds, swept along by fierce winds. We know about these conditions largely from a series of Russian Venera spaceprobes, some of which descended through the atmosphere of Venus in the late 1960s and 1970s; two reached the surface, sending back data briefly before being destroyed by the harsh conditions. The atmosphere of Venus is about 98 per cent carbon dioxide and 2 per cent nitrogen, with traces of a few other gases.

In spite of its complete cloud cover, the surface of Venus has been mapped in great detail using radar from satellites, including some of the Venera orbiters. The latest and best of these surveys was carried out by NASA's Magellan probe, which went into orbit around Venus in August 1990 and mapped almost the entire surface. Although the surface is also heavily cratered, it has much more variation than the surface of Mercury, with a huge plain covering almost two-thirds of the planet (like a dry sea floor) and a mass of land rising above this plain like a continent on Earth. The highest mountains on Venus reach eight kilometres above the surface, and as well as the many impact craters, there are volcanoes, valley systems, and lava flows. But there are far fewer craters on the surface of Venus than on the surfaces of Mercury or the Moon, allowing for the greater area of Venus. By comparing the cratering 'density' on all three bodies, astronomers infer that the entire surface of Venus was renewed about 600 million years ago in some cataclysm that caused lava to flood out from its interior. It may be that the planet has been resurfaced in this way several times in the four billion years or so since it formed, and clearly tectonic activity on Venus is different from the activity that produces continental drift on Earth.

There is another peculiarity about the planet. It rotates very slowly, and in the opposite sense to the rotation of the Sun and most of the planets, taking 243 of our days to turn once. Perhaps this is the result of a major asteroid impact at the end of the massive bombardment that occurred when the Solar System was young. Whatever the cause, it means that, since Venus takes 225 of our days to orbit the Sun once, the combination of this retrograde spin and its orbital motion means that from any point on the equator of Venus the time from noon one day to noon the next would be 116.8 of our days, and there would be just under two such days in every year.

We've already told you a great deal about Planet Earth, the third planet from the Sun in our Solar System, because of its special place as our home. But the Earth – or rather, the Earth–Moon system – is also unique among the planets of the Solar System in another way. Our Moon is about a quarter the size of the Earth, far bigger (in proportion to its planetary parent) than any of the other moons in the Solar System (except for Charon, and we don't regard Pluto as a proper planet, anyway). It has a diameter of 3,476 km, and orbits the Earth at an average distance of 384,400 km (a distance that light takes 1.3 seconds to cover). More than anything else in the Solar System the Moon resembles Mercury, which is 38 per cent the size of the Earth, and a planet in its own right. From the perspective of anyone except an inhabitant of the Earth, it makes much more sense to regard the Earth–Moon system as a double planet, which must have formed in some special way when the Solar System was young. This uniqueness of the Earth–Moon system is even more striking when you note that neither Mercury nor Venus has a moon at all, and the fourth inner planet, Mars, has two tiny moons which are certainly captured asteroids, not primordial companions to the red planet.

So how did the Moon form? With the aid of computer simulations and analysis of rocks brought back from the Moon by the Apollo astronauts, astronomers have developed a compelling model of what went on to make the unique double planet. The Moon, it seems, was ripped out of the Earth in the last stages of the process of planet formation by the impact of an object at least as big as Mars. But this wasn't a process like chipping a piece of solid rock out of a lump of granite. Astronomers refer to the event as 'the Big Splash', and that expression gives you a good feeling for what it was like.

According to this model, the impact of a Mars-sized object with the young Earth generated enough heat to melt the entire surface of the planet to a depth of about a thousand kilometres. The incoming object was completely disrupted in the impact, melted and merged into the ocean of liquid rock that was formed. If the chunk of incoming material had a heavy metallic core, the molten metal would have sunk down through the layers of molten rock to merge with the Earth's iron core; but the rocky parts of the super-asteroid would have merged indistinguishably with the molten rock from the Earth's crust, and some of the resulting mixture would have been thrown up into orbit around the Earth by the Big Splash.

This hot debris would have formed a ring around the Earth, from which all the water and other volatile substances would have evaporated out and been lost into space. But as the material cooled, it would have coalesced, sticking together to form the Moon in exactly the same way that primordial material in a ring around the Sun stuck together to make the planets themselves. The side-effects of the collision have affected the planet ever since – the impact is probably the reason why the Earth spins so rapidly on its axis, so that the day is now just 24 hours long, and an off-centre impact is a likely explanation for the tilt of the Earth, responsible both for the annual cycle of the seasons and for the Milankovitch cycles of the ice ages. We are still directly affected by the impact, four billion years after it happened. And if the Milankovitch cycles really were a decisive factor in turning apemen into human beings, we owe our very existence to the impact.

There is no way to prove the accuracy of this model, short of taking Mars and hurling it into Venus to see what happens. But there is powerful circumstantial evidence in its favour – as well as the strong evidence from the computer simulations of such an impact – from the fact that the Moon does not seem to contain even a trace of water or other volatile substances, and is the only one of the five large, rocky objects in the inner Solar System (Mercury, Venus, Earth, the Moon and Mars) not to have an iron core. Ice recently found on the surface of the Moon is not primordial, but was dumped there by comets.

Mars itself is the first planet you would encounter working outward from the Sun, starting from the Earth–Moon system. Although the distance from Earth to Mars varies according to where the two planets are in their respective orbits around the Sun, at its closest Mars comes within about 56 million km of the Earth, and it has been visited by a fleet of spacecraft which have sent back a wealth of information about the planet, to add to the information gleaned by telescopic studies from Earth.[5]

It orbits around the Sun once every 686.98 of our days, at a distance that varies between 1.38 and 1.67 AU. The day on Mars lasts for 24 hours 37 minutes and 23 seconds, almost the same as the day on Earth, but it has only a thin atmosphere (with just 0.7 per cent of the pressure of Earth's atmosphere at sea level), mostly made of carbon dioxide, and temperatures at the surface range from –140°C to (rarely) just above freezing – in most places temperatures never rise above freezing. The diameter of Mars is 6,795 km (about half that of the Earth); its mass is only a little more than a tenth of the mass of the Earth. Like the other inner planets, its surface is scarred by a mass of craters. In many places, the surface has also been carved by long-gone rivers into a series

[5] Venus comes even closer, within about 42 million km, but only when it is directly between us and the Sun, making it impossible to study through a telescope both because of the Sun's glare and the fact that we would be looking at the night side of the planet, quite apart from the fact that it is covered in clouds!

of canyons and valleys – but it is at least hundreds of millions of years, and probably billions of years, since Mars lost most of its original atmosphere and the planet froze as the greenhouse effect weakened. Any remaining water on Mars is mostly in the form of permafrost, locked away below the surface. Sometimes, some of this frozen water may be temporarily liberated and allowed to flow locally by the heat generated by the impact of an asteroid.

Mars very nearly made it as a planet like the Earth, and clearly would have oceans of liquid water today if it had been as big as the Earth. Venus, equally, would have been quite Earth-like, if only it had started life a little further out from the Sun (or the Sun had been a little cooler) so that liquid water formed and dissolved some of its thick carbon dioxide atmosphere. The intriguing implication is that the 'life zone' around a star like the Sun, the region in which habitable planets might form, extends from a little further out than Venus to at least the orbit of Mars. Instead of regarding the Solar System as being lucky to have one Earth-like planet in its family, it may turn out, when we get to explore other planetary systems, that it is *un*lucky not to have two Earth-like planets!

In our system, though, Mars is distinctly un-Earth-like, and, indeed, un-Venus-like. There are so many craters on Mars that it is clear that what we see today is largely the primordial surface, and that, unlike Venus, it has not been resurfaced by any volcanic cataclysm in the past four billion years. Even so, Mars has been geologically active during its lifetime, and may still be. The largest volcano on Mars is the Olympus Mons, which rises 23 km above the surrounding plains, and has a diameter of 500 km. By comparison, the largest volcano on Earth, Mauna Loa in Hawaii, rises only 9 km above the sea floor, and has a diameter of just 200 km.

As we have mentioned, Mars is accompanied by two tiny 'moons'. Both are lumpy, potato-shaped objects. Phobos, the larger of the two, is about 28 km by 20 km in size and orbits Mars once every 0.3 days at a distance of 9,380 km; the other, Deimos, is about 16 km by 12 km, and orbits Mars every 1.3 days at a distance of 23,460 km. They are both smothered in craters (the largest crater on Phobos is 10 km in diameter, on an object itself only 28 km across at its widest), and can only be bits of debris captured from the nearby asteroid belt by the gravity of Mars. That brings us nicely on to the next feature of the Solar System – the asteroid belt itself.

The asteroids are small, rocky lumps of debris (much smaller than planets), many of which orbit in a band between the orbits of Mars and Jupiter. Because they are so small, the asteroids reflect very little light from the Sun, and the first members of the asteroid belt were only discovered in the nineteenth century; now, more than 2,500 of these objects have been identified and catalogued, orbiting at distances between 2.2 AU and 3.3 AU from the Sun. Hundreds more have been seen intermittently, but not for long enough for their orbits to be calculated accurately; there may be as many as half a million objects in the asteroid belt large enough to be photographed by the 200-inch telescope on Mount Palomar. But out of all these bits of rubble, only five have diameters greater than 300 km, and only about 250 are more than 100 km across. Most of the known asteroids are about a kilometre across. The total mass of all the objects in the asteroid belt put together is only just over a quarter of the mass of the Moon; they take between three and six years to orbit the Sun once, depending on their exact distance from the Sun.

The reason why all of this debris failed to stick together to form a single large object is that it is constantly being disturbed by the gravitational pull of Jupiter, which tugged the primordial fragments into a confusion of orbits in which they were likely to smash into each other violently, instead of snuggling up and sticking together. In the region now occupied by the asteroid belt, the models suggest that there should have been enough primordial material to make four rocky planets the size of the Earth (or one planet with four times the mass of the Earth). Some of the meteorites that reach the Earth, and are thought to be originally from the asteroid belt, contain metallic material, showing that they were indeed once in the cores of larger bodies.

One favoured model today (alas, there is no way to prove it) is that the original material in the asteroid belt actually formed about eight objects the size of Mars, and that most of these broke up, over an interval of about a hundred million years, as a result of collisions caused by the disturbing influence of Jupiter. One of the super-asteroids was disturbed sufficiently that it fell into the inner Solar System, colliding with the Earth and producing the Moon in the Big Splash. One of the objects is still there – Mars itself. The rest got broken up in collisions and scattered, with most of the debris produced getting flung out of the asteroid belt entirely, either into orbits taking it to a fiery end in the Sun or out of the Solar System altogether.

The importance of this is obvious. Although the remaining asteroids are in relatively stable orbits (because all the ones in unstable orbits have long since been ejected), Jupiter is still at work, tugging on them gravitationally. Collisions between asteroids still occur, and fragments from those collisions are still sometimes perturbed into orbits which take them across the orbit of Earth. The rain of cosmic debris which scarred the faces of the inner planets when the Solar System was young has not yet stopped, merely slowed to a trickle – and the impact of a lump of debris ten kilometres or so across with the Earth was enough to bring an end to the era of the dinosaurs. Fortunately, such events are now rare; but in the very long term, if civilisation is to survive on Earth it will have to find a way to protect itself from cosmic impacts.

The step beyond the asteroid belt to the next member of the Sun's family provides the greatest contrast in the Solar System – from planetary pebbles a kilometre or so across to the largest planet in the Solar System, Jupiter, with a diameter of 143,000 km (one tenth of the diameter of the Sun) and a mass 318 times the mass of the Earth (a full 0.1 per cent of the mass of the Sun). It is this enormous mass (for a planet) that makes Jupiter so important as a gravitational influence on the rest of the Solar System, and it owes its mass to the fact that it formed far enough out from the Sun to retain vast amounts of primordial gas – 90 per cent of Jupiter is hydrogen, 10 per cent helium, with only traces of gases like methane and ammonia. It orbits at an average distance of 5.2 AU from the Sun, taking 11.86 years to complete one orbit.

Everything about Jupiter involves superlatives. It is striped by coloured bands produced by circulation in the atmosphere, the equivalent of the jet streams that blow around the Earth at high altitudes. But a single storm on Jupiter, the Great Red Spot, has been swirling around the planet for at least three hundred years, and is big enough to swallow the entire Earth.

The most impressive thing about Jupiter, though, is that it has its own family of moons – a Solar System is miniature. There are four large jovian satellites (Io, Europa, Ganymede and Callisto), which were discovered by Galileo at the beginning of the seventeenth century. This discovery of satellites in orbit around their parent planet helped to establish the idea that the Earth orbits the Sun in a similar way, shattering centuries of dogma that the Earth lay at the centre of the Universe. Jupiter also has at least a dozen smaller satellites (the number keeps going up as more are discovered), many of which are captured asteroids.

The four galilean satellites are remarkable objects in themselves. Io has been shown by spaceprobes to be a shining ball of red and orange sulphurous material, which pours out of active volcanoes that dot the surface of the moon. The heat that powers the volcanoes comes from the tidal forces of Jupiter, which squeeze the interior of the moon in their grip. Europa, by contrast, is entirely covered in ice, which is criss-crossed by dark cracks. Also warmed by the tidal squeezing of Jupiter, Europa has a sea of liquid water beneath its crust of ice, and may provide a home for life. Callisto, however, is too far out from Jupiter to be heated much by tidal squeezing, and is covered with a thick layer of solid ice, scarred by the most heavy cratering seen on any object in the Solar System. And Ganymede, the largest satellite in the Solar System, has an ice surface which is partly cratered, like Callisto, but partly smooth, apparently covered by fresh ice which must have been spread over the surface in some relatively recent cataclysm.

After Jupiter, the remaining planets in the Solar System are something of an anticlimax. Saturn is a smaller Jupiter, with a diameter only 9.4 times that of the Earth and a mass 95 times that of the Earth, orbiting the Sun once every 29.46 years at a distance varying between 9 AU and 10 AU. But the Saturn system is distinguished by two features – the famous rings, which give it the most beautiful appearance of any of the outer planets in astronomical photographs, and a family of satellites which includes Titan, the most intriguing moon in the Solar System. Titan has a diameter of 5,150 km, making it a little smaller than Ganymede (diameter 5,262 km). The special thing about Titan, though, is that it has a thick atmosphere, mainly made of nitrogen but also rich in methane. The surface pressure on Titan is 1.6 times the atmospheric pressure at sea level on Earth, and the temperature there is –180°C. There may be lakes or oceans of liquid methane on the surface, with methane rain falling from the clouds. Titan is like a frozen, smaller version of the early Earth. When the Sun nears the end of its life and swells up to become a red giant, the inner planets, including the Earth, will be fried to a crisp, but Titan could be warmed sufficiently to allow life a second chance in the Solar System (a third chance, if Europa really is another home for life). But that lies far in the future. More immediately, the Cassini spacecraft, launched at the end of 1997, will drop a probe into the atmosphere of Titan early in the twenty-first century, in the hope that studying its chilly atmosphere may provide clues to what the Earth's atmosphere was like long ago, and perhaps even insights into the origin of life.

Beyond Saturn there are two more gas giants. Uranus orbits the Sun once every 84.01 years, at a distance varying between 18.31 AU and 20.07 AU. But it has a mass only 14.5 times that of the Earth, and a diameter four times that of the Earth. Neptune, the most distant real planet in the Sun's family, orbits the Sun once every 164.79 years, at a distance of 30.06 AU. Its mass is a little more impressive than that of Uranus, 17.2 times the mass of the Earth, but its diameter is marginally (about 1 per cent) less than that of Uranus. The last of the giant planets lies only about thirty times as far from the Sun as we do, and in comparison with the distances between the stars, the entire family of planets huddles close around the central solar fire. In terms of the time it takes light to travel across space, the distance from Neptune to the Sun is just under 4.2 light hours; but the distance from the Sun to the nearest other star is just over 4.2 light *years* – nearly nine thousand times further. There is, though, something between Neptune and the nearest star, and that something is still – just – part of the Sun's family.

Beyond the orbit of Jupiter, the heat of the Sun is so feeble that debris left over from the formation of the Solar System can still exist in the form of icy lumps, not just as rocky asteroids (although the icy lumps may well have rocks embedded in them). These cosmic icebergs contain much more than just frozen water – solid carbon dioxide, methane, and ammonia are all present in them. Sometimes these icebergs get disturbed into orbits which carry them in close to the Sun, before they swing round the Sun and back out into the depths of space. As they approach the Sun and warm up, some of the ice evaporates and forms a long tail, which shines by reflected sunlight; the icy lump has become a comet. But when it leaves the inner Solar System, the tail fades away as the heat from the Sun wanes, and the nucleus of the comet becomes an inert, icy lump once again.

Studies of the orbits of comets show that their ultimate origin is a spherical cloud of icebergs far out in the depths of space, which surrounds the Sun, literally halfway to the nearest star, at a distance of about 100,000 AU, or a couple of light years. A typical comet nucleus in this cloud orbits around the Sun at about a hundred metres a second, and may have been there for billions of years, since the Solar System formed. But occasionally some outside influence, such as the gravitational tug of a passing star, may send some of these icy lumps falling into the inner part of the Solar System, accelerating all the way but taking millions of years on the journey before whipping round the Sun and heading back out into space. Some of these visitors to the inner Solar System are captured by the gravity of Jupiter into shorter, but still elongated orbits, and, like Halley's Comet, make repeated passes by the Sun every few decades or centuries, before they evaporate entirely, leaving odd lumps of rock, grit and dust scattered along their orbits.

When the Earth passes through such a comet trail, the sky is lit up by the bright streaks of meteors, each one caused by a piece of cosmic dust no bigger than a grain of sand burning up in the atmosphere. But if a comet strikes the Earth, it can do as much damage as a rocky asteroid – probably more, since it is likely to be travelling much faster, having fallen in from so far away, and will therefore carry more kinetic energy than an asteroid of the same mass. Indeed, it is most likely that the impact that caused the death of the dinosaurs was a comet, not an asteroid.

It was largely thanks to Jupiter that the comets got where they are today, out on the fringes of the Solar System. When the planets formed, there must have been huge numbers of these cosmic icebergs in the region between Jupiter and Neptune, but, like the primordial debris of the asteroid belt, under the gravitational influence of Jupiter and the other giant planets they would have been disturbed into orbits either taking them to their doom in the Sun or out to the fringes of interstellar space. There is still a belt of this kind of icy debris beyond the orbit of Neptune. These objects are only just being detected by the latest telescopes, but on the evidence so far it seems that there may be a billion comets in this Kuiper Belt. The belt flares outwards (in cross-section, rather like a huge trumpet) to link up with the spherical Oort Cloud of comets in deep space, and there must be about ten trillion comets in all. The entire mass of all the comets put together is only a few times the mass of the Earth, but the largest cosmic icebergs detected so far, between the orbits of the outer planets, are a couple of hundred kilometres in diameter (about a tenth of the size of Pluto; Pluto now seems to be just a large example of this kind of iceberg).

Any object in this part of the Solar System will eventually be perturbed by Jupiter's gravitational influence, either coming into our part of the Solar System or out in to deep space (and it cannot have been in that orbit for more than a few million years, having presumably worked its way in from further out in the Solar System). If it comes inward, it is likely to fragment as it heats up, and gases boiling off from it will crack the ice, producing a swarm of comets passing through the inner Solar System together. Even if none of them were to strike the Earth, the amount of fine dust spread round the inner Solar System as the comets evaporated could be enough to attenuate the amount of heat reaching the Earth from the Sun – some astronomers believe that this could be the trigger for some ice ages on Earth, a kind of cosmic winter. If so, the course of evolution and civilisation may be directly influenced by the stars, although not in the way that astrologers think. A passing star could shake a supercomet loose from the

Oort Cloud, sending it tumbling inwards across the Solar System to break up near the Sun and cause an ice age on Earth millions of years later.

Any such suggestions must be speculative, and we may never know how accurate these scenarios are. But they serve to remind us that the Sun and its family do not exist in isolation, but are part of a much bigger star system, the Milky Way Galaxy. To put the Solar System in perspective, we need to look at the lives of the stars themselves – which, since we are moving to more extreme conditions of temperature and pressure, means that the physics gets that much simpler to deal with. But you may find that the timescales and distances involved in this part of the story take some getting used to.

CHAPTER TEN
THE LIVES OF THE STARS

One of the most fantastic claims in the whole of science is that we have good models to describe what goes on inside stars – how they are born, how they live, and how they die. Stars appear to us as no more than points of light on the sky, at distances so enormous that light takes hundreds or even thousands of years to travel from them to us (even the fact that we know the distances to the stars accurately is an achievement that would have amazed astronomers less than two hundred years ago). Although atoms and subatomic particles are, in a sense, just as remote from everyday experience, because we cannot see them with our own eyes, at least they exist in laboratories down here on Earth, and can be directly studied as the various models are tested in experiments; there is no way to test our models of stars by carrying out experiments on such distant objects.

Even if we could, we might not live to see the outcome of those experiments – not because they would be dangerous, but because old age would catch up with us first. Those stellar models tell us that in many cases stars live for billions of years. A human lifespan is usually less than a hundred years, and the lifetime of our entire civilisation is less than a hundred human lifetimes – less than 10,000 years. How can anyone seriously claim to know how a star was born, billions of years ago, and how it will end its life, billions of years hence?

In fact, the models that describe stellar structure and evolution[1]

[1] Astronomers traditionally use the word 'evolution' to refer to the life cycle of a single object, such as a star or galaxy – the way it changes as it gets older. They are not referring to Darwinian evolution, with one variety of star or galaxy being replaced by another cosmic species.

are among the greatest triumphs of modern science, and have been tested to high accuracy. They involve a combination of observations of the stars themselves (using the techniques of spectroscopy), computer simulations of what goes on inside stars (based on the known laws of physics), and even actual experiments, carried out here on Earth, to test aspects of the models, notably the rate at which certain nuclear reactions go on under conditions corresponding to those predicted to exist at the heart of a star by rival models. The whole package hangs together scientifically. But the key ingredient, the one without which astronomy would be no more than a kind of cosmic bean-counting, is spectroscopy.

As we mentioned in chapter two, spectroscopy enables astrophysicists to identify the different chemical elements present at the surface of a star, by analysing the light from that star – it even enabled Norman Lockyer to identify a previously unknown element, helium, from its spectral signature in sunlight. The distribution of starlight among photons with different energies (different colours) in the light from a star also tells us the temperature at the surface, using the famous black-body energy curve which was so important in setting Max Planck on the path to quantum physics.

If you know the temperature at the surface of a star and its mass, then the basic laws of physics (the relationships between properties such as the temperature and pressure of a gas) and the computer models tell you how hot the star must be inside, what the pressure and density must be, and so on (we shall explain how we know the masses of stars shortly). And if you know the temperature and pressure at the heart of a star, and you know what the star is made of, then you know what nuclear reactions are going on inside the star. Experiments in laboratories on Earth then tell you how much energy those reactions ought to be producing. This can be compared with the amount of energy a star actually does radiate into space, and the models can be tweaked to make theory and observation agree more closely with one another. It all fits together beautifully, and uses an enormous amount of physics, ranging from things so standard that no physicist really thinks much about them (such as the temperature/pressure relationship) to the most sophisticated nuclear physics experiments. The success of astrophysics is in many ways the culmination of the scientific way of doing things, confirming that all of the bits of physics that were discovered separately really do work in the way that our models suggest when they are all put together. And all of this is most easily seen, of course, for the nearest star – the Sun.

The birth of astrophysics can be dated quite precisely, to a talk given in 1920 by the pioneering astrophysicist Arthur Eddington at the annual meeting of the British Association for the Advancement of Science, held that year in August, in Cardiff. Until the beginning of the twentieth century, it had been a major puzzle for astronomers to explain where the Sun got its energy from. Geological evidence, and Darwin's theory of evolution by natural selection, required a very long history for the Earth, and, by implication, for the Sun. But no known form of chemical energy (such as burning coal) could explain how the Sun could have stayed hot for long enough for geology and evolution to do their work.

The discovery of radioactivity, involving energy released from the nuclei of atoms, began to change the picture, and Albert Einstein quantified the enormity of that change with his famous equation which tells us that matter itself can be converted into energy. But at first people were reluctant to accept the full implications of these new ideas. It was one thing to use the idea of sub-atomic (that is, nuclear) reactions to explain why a piece of radium felt warm to the touch; but an enormous leap of faith to accept that the same kind of process might explain the vast outpouring of energy from the Sun, equivalent (according to Einstein's equation) to converting five million tonnes of matter into pure energy every second.

For a time, people clung to a nineteenth-century idea that a star like the Sun might keep hot for long enough to explain events on Earth simply by shrinking very slowly under its own weight, turning gravitational potential energy into heat as it did so. But even that could only keep it shining for a few tens of millions of years. It was Eddington who finally squashed this idea, and put astrophysics on the right track, telling his colleagues in Cardiff that,

> Only the inertia of tradition keeps the contraction hypothesis alive – or rather, not alive, but an unburied corpse. But if we decide to inter the corpse, let us freely recognise the position in which we are left. A star is drawing on some vast reservoir of energy by means unknown to us. This reservoir can scarcely be other than the sub-atomic energy which, it is known, exists abundantly in all matter; we sometimes dream that man will one day learn to release it and use it for his service. The store is well-nigh inexhaustible, if only it could be tapped. There is sufficient in the Sun to maintain its output of heat for 15 billion years . . .

Over the next few years Eddington demonstrated how even the simplest laws of physics could be applied to reveal deep truths about the nature of stars. If a star is in equilibrium – neither expanding nor contracting – the inward pull of gravity must be balanced by the outward pressure of hot material inside the star. The laws of physics tell us that a ball of gas containing a certain amount of matter and held up by the pressure inside must have a certain size, a certain central temperature, and radiate a certain amount of energy – *wherever* that energy comes from. The mass of the Sun is known quite precisely, from the way its gravity influences the orbits of the planets. The masses of many stars can be calculated in a similar way because very many stars have other stars as companions, forming binary systems. If binary stars are close enough to us for us to measure the orbital properties of the two stars as they move around one another (spectroscopy comes in here once again, because the Doppler shift in the spectral lines from the stars tells us how fast they are moving in their orbits) then their masses can also be calculated.

The amount of energy being put out by a star can also be estimated, by measuring its apparent brightness on the sky and making allowance for how far away it is. This is where the need to measure distances comes in.

It was only at the end of the 1830s that the first accurate distances to the nearest stars were measured. The technique used was triangulation – exactly the same technique used by surveyors on Earth, when a distant object is looked at from the two ends of a long, carefully measured baseline, and the angles between the two lines of sight and the baseline used to calculate the distance to the object from our knowledge of the properties of triangles. The baseline used in astronomical surveying of this kind is the diameter of the orbit of the Earth around the Sun.[2]

Observations are made six months apart when the Earth is on opposite sides of its orbit, and just a few nearby stars are close enough that they seem to move slightly against the background of more distant stars over this interval. The effect is known as parallax, and is exactly like the way in which if you hold one finger up at arm's length and close each of your eyes in turn the finger seems to move against the distant background. Because angles are measured in seconds of arc, this technique gives distances measured in terms of parallax seconds of arc, or parsecs. One parsec is just over 3.25 light years, rather more than 206,000 times the distance from the Earth to the Sun.

The first stars to have their distances measured, just over a century and a half ago, were 61 Cygni (3.4 parsecs, or just over 11 light years, away), Alpha Lyrae (8.3 parsecs, or 27 light years), and Alpha Centauri, which is now known to be a multiple star system, which includes the closest star to the Sun, at a distance of 1.3 parsecs (4.3 light years). The nearest star system to the Sun is seven thousand times further away than the average distance of Pluto from the Sun. In the past decade, distances to many somewhat less nearby stars have been measured with exquisite precision, using this technique, by the Hipparcos satellite, orbiting clear of the obscuration caused by the Earth's atmosphere. Other geometrical techniques, taking advantage of the way stars at still greater distances move across the sky, have also been used down the years, and have revealed the distances to stars and clusters of stars. And these techniques have together proved just good enough to provide information about distances to star clusters that contain the most important astronomical stepping stones: stars known as Cepheid variables.

The great thing about Cepheids is that they vary in brightness in a very regular way, and that for any particular Cepheid the

[2] And we know the diameter of the Earth's orbit by applying similar triangulation techniques across the Solar System.

period of the variation (the time it takes to brighten, dim, and brighten again) depends on the actual average brightness, or luminosity, of the star. So if you measure the period of a Cepheid and its apparent brightness on the sky, you can calculate how far away it is by comparing the apparent brightness with the intrinsic brightness determined from the period–luminosity relationship – just as you could determine the distance to a 100-watt lightbulb by measuring its apparent brightness. The fainter it looks the further away it must be.

The technique *just* works; the distances were known to scarcely a handful of Cepheids before the Hipparcos data became available, and it was these few stars that were used to calibrate the period–luminosity relationship. Once it was calibrated it was possible to use the relationship to give the distances to Cepheids in clusters right across the Milky Way, and even in our nearer neighbour galaxies (more of this in the next chapter). And that is how we know that the Milky Way Galaxy is a flattened disc about 4 kiloparsecs (13 thousand light years) thick in the middle, some 30 kiloparsecs (98 thousand light years) across, embedded in a spherical halo of globular star clusters that extends across a volume of space 150 kiloparsecs (490 thousand light years) in diameter. The Sun is a very ordinary star, situated in an ordinary part of this Galaxy, about 9 kiloparsecs out from the centre of the disc, roughly two-thirds of the way from the centre of the Milky Way to the edge of the disc, which has a radius of 15 kiloparsecs.

We understand the lives of the stars in that disc because we can (at least in some cases) measure their distances and brightnesses, determine their masses, and study their composition using spectroscopy. We also have one other invaluable tool – statistics. There are so many stars to study that we can see them at all stages in their life cycles, and can compare their observed properties with the predictions of our computer models – the astronomical equivalent of testing models by experiment. In exactly the same way, you could work out the life cycle of a tree by studying a single forest for a few weeks, and investigating trees in all stages of their life, without having to watch long enough to see an individual seedling grow to become a mature tree and produce seedlings in its turn.

The laws of physics that have to be applied to describe what goes on under the conditions that exist inside a star are particularly simple (much simpler than the laws that describe what goes on inside a tree) because, as we have explained, electrons are stripped from their atoms under these conditions of high temperature and pressure, so that the individual atomic nuclei behave like the components of a perfect gas. Once any ball of gas contains enough mass for this to happen, forming a so-called plasma of charged particles, fast-moving charged particles convert some of their kinetic energy into radiation, and this radiation in turn interacts with other charged particles, contributing to the pressure that holds the ball of gas up against its own weight. A glowing, stable star is held up by a combination of gas pressure and radiation pressure. But if the ball of gas has more than a certain amount of mass the conditions in its heart become so extreme that the fast-moving particles radiate huge amounts of electromagnetic radiation – energy which blows the star apart.

Eddington realised that there are three, and only three, possible fates for a ball of gas in space that collapses under its own weight. If it is too small for a plasma to form in its heart as electrons are stripped from atoms, it will become a cool globe held up only by gas pressure – something like Saturn. If it is rather bigger, it can become a glowing star, held up by a mixture of gas pressure and radiation pressure. And if it is bigger than a certain size, it will briefly shine as a superhot gas globe, before being blown apart by radiation pressure. From this simple piece of physics, Eddington worked out just what range of masses allow stable stars to exist. The exact numbers you get out of these calculations depend on the actual composition assumed for the model star, because the number of electrons available depends on which atoms (nuclei) are present (one electron for each hydrogen atom, eight for each oxygen nucleus, and so on). For this reason, Eddington's numbers don't quite match up with modern versions of the calculation. But the broad details remain unaffected even though we have a better idea today of what stars are made of than Eddington had in the 1920s.

Eddington asked us to imagine a series of globes of gas of various sizes, starting with 10 grams, then 100 grams, 1,000 grams, and so on. The *n*th globe has a mass of 10^n grams. It turns out that according to the basic laws of physics the *only* globes for which the basic laws of physics allow a stable existence as a glowing star are numbers 32 to 35 in the series. So what happens if we test this prediction from basic physics with observations of the real Universe? Globe 31 in the sequence has a mass of 10^{31} grams, which is roughly five times the mass of Jupiter. Sure enough, Jupiter is a ball of gas held up by gas pressure, not a glowing star. Globe 32 has a mass of 10^{32} grams, about one-tenth of the mass of the Sun. So a star cannot begin to glow until it is several times bigger than Jupiter and about a tenth as massive as the Sun – the Sun does indeed sit on the correct side of the dividing line.[3] And at the other extreme, the actual masses of the largest known stars are no more than a few times 10^{35} grams, about a hundred times the mass of the Sun. Globe 35 really is the biggest one that can form a stable star.

The other key fact that emerged from Eddington's investigation is that all stars, regardless of their mass and brightness, must have much the same temperature in their hearts. He calculated this temperature to be about 40 million degrees Celsius because he allowed for the presence of rather too many electrons; the modern version of the calculation reduces this estimate to 15–20 million degrees. In his great book *The Internal Constitution of the Stars* (CUP, 1926) Eddington summed things up – just *before* the new quantum physics became established. Using examples from specific stars whose mass and brightness were known, he pointed out that by applying the simple laws of physics:

> Taken at face value ... whether [an energy] supply of 680 ergs per gram is needed (V Puppis) or whether a supply of 0.08 ergs per gram

[3] The number of atoms of hydrogen you would need to put together to make a star with the minimum amount of mass is about 10^{57}, a number remembered by astronomers as 'the Heinz soup parameter', even though, alas, we are told that there never really were '57 varieties' of Heinz products.

is needed (Krueger 60) the star has to rise to 40,000,000° to get it. At this temperature it taps an unlimited supply.

The reason why all stars have much the same internal temperature is thanks to feedback. If a star (any star) shrinks a little, it will get hotter inside as more gravitational energy is released. That would make it expand, restoring equilibrium. Or suppose the star expanded anyway. That would take energy out of its heart, making it cool a little, so that the pressure fell and it contracted once again. Stars have an inbuilt thermostat which keeps their cores at just the right temperature for subatomic (nuclear) energy to be released.

Incidentally, this feedback helps to explain a question which generations of astronomy lecturers have taxed their new students with. The question asks what effect nuclear reactions have on the temperature at the heart of a star. The obvious answer is that nuclear reactions keep a star hot – but it is wrong. Without those nuclear reactions to produce energy and thereby an outward pressure to resist the inward tug of gravity, the star would shrink and get hotter still! The role of nuclear reactions is to keep the heart of a star *cool*, relatively speaking.

In the early 1920s, physicists dismissed Eddington's whole scenario as absurd. The reason was that they knew how much kinetic energy two protons would need in order to collide strongly enough to overcome their mutual electrical repulsion and fuse, making helium, and that energy would require temperatures much higher than Eddington claimed for the temperature at the heart of a star. Famously, Eddington stuck to his guns, pointing out that 'the helium which we handle must have been put together at some time and some place' and going on, tongue in cheek, to say that 'we do not argue with the critic who urges that the stars are not hot enough for this process; we tell him to go and find *a hotter place*'. This is usually interpreted as Eddington's way of telling his critics to go to Hell. And even as the book in which those words appeared was being printed, physics was being transformed by the quantum revolution.

Within a couple of years, George Gamow's application of quantum uncertainty to describe the fusion of atomic nuclei had explained just how protons really could merge together at the temperatures that existed at the hearts of stars. It was a triumph for astrophysics, which established this branch of science and made it one of the most exciting areas of investigation for the next forty years. We shan't go into all the details of that investigation; but we hope the example we have given of the power of physics in explaining what goes on inside so simple a thing as a star will convince you that astrophysicists really do know what they are talking about when they describe the lives of the stars. So – let's look at this life cycle in detail, basing our account on all the best information gleaned since Eddington's day, and the latest and best computer simulations.

In a galaxy like the Milky Way today, star formation is a continuous process that involves recycling material from clouds of gas and dust in space. Hot young stars are found embedded in clouds of material, orbited by dusty discs (in which planets will form) spinning around the young stars and with jets of material being ejected from the young stars, blown outward from their polar regions by the pressure of the radiation from the young star itself. The clouds from which stars form are cool enough for stable molecules, including things like carbon monoxide, to exist in them, and are known as molecular clouds; they lie in the plane of the Milky Way, and are each a few light years across, with densities of no more than 1,000 to 10,000 molecules per cubic centimetre. These clouds are made almost entirely of hydrogen (75 per cent) and helium (25 per cent), with just a smattering of heavier elements. Star formation occurs when such a cloud gets squeezed from outside, and starts to collapse, to the point where its own weight makes it continue collapsing, fragmenting as it does so to form individual stars, binary stars, and more complicated systems. This collapse, and the whole process of star formation today, seems to be intimately linked to the spiral structure of our Galaxy (there is another family of galaxies, called ellipticals, which do not have this spiral structure, and in which star formation does not seem to be occurring today).

The stars in the disc of the Milky Way, as in many other disc-shaped galaxies, form a spiral pattern, rather like the pattern made by cream being stirred into a cup of coffee. This pattern is caused by a compression wave moving round the Galaxy, rather like a sound wave. A sound wave squeezes the air as it passes, but leaves individual molecules of air more or less in their original places after it has passed. In the same way, the galactic wave squeezes the stars and clouds of the Milky Way as it passes, but leaves them in much the same places after it has gone. The squeezing process itself triggers the collapse of some clouds of interstellar material, and the larger of the stars that form as a result run through their life cycles very quickly (in no more than a few million years) and explode as supernovas (more of them later). Because such large stars are so short-lived, there has not been time for them to move far away from where they were born, and these explosions also happen close to the region where the galactic wave is doing its work. The blast from these exploding stars triggers the collapse of more clouds in that part of the Galaxy. So although the density wave that moves around the Galaxy is itself invisible, it is edged by a distinctive border of hot, bright young stars, born as a result of the passing of the wave.

The classic example of such a stellar nursery visible in the night sky from the Earth is in the constellation Orion. There is a cloud of gas and dust in this constellation, known as the Orion Nebula, which is illuminated by the light from the young stars embedded in the nebula. In our part of the Galaxy, stars are separated by distances of a few parsecs, and there is roughly one star for every five cubic parsecs of space; in the Orion nursery, the stars are only one-tenth of a parsec apart, and there are ten thousand stars per cubic parsec. This is typical of the way stars are born, although they get spread out as they each move in their own orbits around the centre of the Milky Way. But we shall describe what happens to just a single star after it is born.

The key problem that a collapsing cloud has to deal with, if it is to form a star, is to get rid of angular momentum. This is why discs of material form around young stars, and planets form in the discs. Angular momentum is also removed as material is blown away into space. Often, the core of the collapsing cloud will split into two components, two proto-stars orbiting around each other at a respectable distance; this also stores angular momentum in a way that doesn't stop the individual stars collapsing. And it is by no means unusual for each of these two cores to split in their turn, forming a double binary system.

A star like the Sun begins to operate as a nuclear fusion reactor when the temperature in its core rises to about 15 million degrees. Then, as long as there is hydrogen fuel available in the core to turn into helium, the nuclear reactions stop the star from collapsing any further and getting any hotter inside. The lifetime of a star in this stable, hydrogen-burning state (known as the 'main sequence') depends on its mass – but the bigger a star is, the less time it spends as a main-sequence star, because it has to burn its fuel more vigorously to hold itself up against its own weight. It does so not by being much hotter inside, but by burning its fuel more rapidly at the same sort of temperature. A star twenty-five times as massive as the Sun spends only three million years in this state; the Sun itself will be a main-sequence star for a total of about 10,000 million years; and a star half as massive as the Sun can sit quietly turning hydrogen into helium for 200,000 million years. The Sun is now almost exactly halfway through its life as a main-sequence star.

When all the hydrogen in the core of the Sun has been converted into helium, it will still be surrounded by an 'atmosphere' of hydrogen, containing about half the original mass of the Sun. With its energy supply exhausted, the core of the star will begin to shrink, generating heat as gravitational energy is released. This heat will make the outer part of the star expand, turning it into a red giant star, and will also trigger nuclear hydrogen burning at the base of the atmosphere, encouraging the expansion. A lot of material is lost into space at this stage of the life of a star, blown away by this process. But even allowing for the amount of matter lost altogether when this happens to the Sun, it will still swell up to engulf the orbit of Mercury, almost reaching to the present-day orbit of Venus. The Sun will lose so much mass in the process, though, that the orbits of the planets will expand, as the Sun loosens its gravitational grip, so Venus will not be engulfed, merely fried.

Before long, while all this is going on the temperature in the collapsing helium core of the star will rise to around 100 million degrees Celsius, and a new phase of nuclear burning will begin. In this process, helium is converted into carbon, and energy is released. So the star stabilises once again as a red giant. This phase of its life is very short-lived – in the case of the Sun, it will last for only about a billion years.

At the end of this breathing space, all the helium is exhausted, and the carbon core begins to collapse, releasing enough heat to encourage a further wave of hydrogen burning further out from the centre, with the star expanding to reach the present orbit of the Earth. In this phase of its life, the star becomes very unstable, with its atmosphere puffing itself up and losing material into space, then shrinking down again, repeatedly. This helps to seed interstellar clouds with elements such as carbon and nitrogen. Eventually, all that is left is the cooling inner core, itself made mainly of carbon, no longer collapsing but held up by the physical pressure of the nuclei and electrons jostling against one another.

Ultimately, the limiting density is reached, and the collapse stops, when quantum forces stop the electrons from being squeezed even more closely together (remember that no two electrons can occupy the same quantum state – it is the same quantum exclusion that stops atoms collapsing that holds a so-called 'degenerate star' up against its own weight). This only happens when the stellar remnant, typically with two-thirds as much mass as the Sun has today, has shrunk to about the size of the Earth, becoming what is known as a white dwarf.[4] One cubic centimetre of white dwarf material would have a mass of about a tonne – it would have a million times the density of water.

Stars that start out rather more massive than the Sun can go through more stages of nuclear burning, at successively higher temperatures, manufacturing elements such as oxygen and neon out of carbon, and spreading them into the interstellar clouds. Except for hydrogen and helium, every element in the Universe, including every element on Earth and in your body, has been

[4] To put this in perspective, remember that today the Sun has about a million times the volume of the Earth.

manufactured inside a star. But a star which starts out with more than about eleven times as much mass as the Sun is still left, even after all this mass loss, with well over a solar mass of material in its leftover stellar ember. At that point things start to get even more interesting.

If a dead star, no longer held up by nuclear burning, has more than about 1.4 times the mass of our Sun, then even quantum processes cannot stop a further stage of compression. What happens is that the electrons which are roaming about among the atomic nuclei in a white dwarf are forced to combine with the protons in those nuclei, to make neutrons. This is inverse beta decay, the reverse of the process that occurs on Earth today when an isolated neutron spontaneously converts itself into an electron and a proton (plus an anti-neutrino). It reduces the stellar cinder to nothing more than a ball of neutrons, which pack together in the most efficient form that matter can be packed together, essentially forming a single atomic nucleus with the mass of a star like the Sun. Whereas a white dwarf with about the mass of the Sun is the size of the Earth, a neutron star with one and a half times the mass of the Sun is only about ten kilometres across, the size of a large mountain on Earth. The collapse of a stellar core from the white dwarf state (or some even less compact state) to a neutron star would release an enormous amount of gravitational energy, with a great deal of it carried away by neutrinos, since one neutrino is released every time a proton and an electron merge to form a neutron. Each cubic centimetre of the material in a neutron star would weigh about a billion tonnes.

Even this is not the end of the story of stellar collapse. It is quantum processes that hold a neutron star up against the crushing inward pull of gravity, but these quantum processes themselves are limited and gravity is not. A compact ball of matter with only three or more times as much mass as our Sun, no longer held up by the energy released by nuclear burning, cannot hold itself up at all. No neutron star with more than this mass can exist (according to theory; and, comfortingly for theorists, observations of real stars have yet to show the existence of any neutron star with more than this much mass, just as they have yet to show the existence of any white dwarf star with more than 1.4 times the Sun's mass). For such an object, gravity completely overwhelms even quantum forces, and the stellar cinder shrinks indefinitely, down towards a point of infinite density, a singularity. On the way, it becomes so dense that its gravitational field becomes so strong that nothing can escape from it, not even light. It becomes a black hole. Neutron stars and black holes are made in our Galaxy today when stars with more than eleven solar masses of material reach the end of their lives. It happens like this.

The problem with using nuclear fusion to provide the energy to hold a star up is that there are limits to what it can do. Specifically, once you have made nuclei of the iron family (iron itself, nickel and the like) it takes *more* energy to make still heavier nuclei, and you cannot get any more energy out by the fusion process. You have to put energy *in* to make the really heavy elements, such as platinum, mercury, gold or uranium. This is because nuclei of iron-56 represent the most energetically stable configuration possible for protons and neutrons in the form of an atomic nucleus.

Using the familiar analogy, it's as if the iron family nuclei sit at the bottom of a valley, with the lighter elements up one side of the valley and the heavier elements up the other side of the valley. All of the lighter elements would 'like' to be at the bottom of the valley, in the low energy state. Lighter ones can achieve this, in principle, by fusion (provided the nuclei have enough thermal energy to overcome their mutual electrical repulsion); heavier ones can achieve this by fission (although very many heavy elements are relatively stable once they have formed, as if they sat in deep potholes on the side of the valley). But it is still true that the heavier elements are made inside stars in the first place, in processes far more energetic than anything going on inside a star like the Sun.

We mentioned that nuclear burning inside stars begins with the conversion of four protons (by a slightly circuitous route) into one alpha particle, a nucleus of helium, made up of two protons and two neutrons (along the way, two of the original protons are converted into neutrons by inverse beta decay). The helium nucleus is itself a particularly stable unit, and later stages of nuclear fusion inside stars basically involve sticking alpha particles together.

It happens that the seemingly obvious first step in this process, sticking two alpha particles together to make a single nucleus of beryllium-8 (four protons plus four neutrons), doesn't work – beryllium-8 is extremely unstable, and promptly splits into two alpha particles. So the 'ash' produced by helium burning is composed of nuclei of carbon, each made up of three alpha particles. Carbon-12 nuclei form in relatively rare interactions involving three helium nuclei at once. From then on, though, if a star is massive enough and gets hot enough inside at each stage of its collapse, the elements are built up largely by sticking alpha particles, one at a time, on to existing nuclei. This is why common elements such as carbon (6 protons and 6 neutrons, making 12 nucleons in all), and oxygen (16 nucleons) have the nuclear structure they do. Sometimes this is followed by the absorption of another proton (or two), and maybe the ejection of a positron, converting one of those protons into a neutron, making nuclei such as nitrogen (7 protons, 7 neutrons).

If a star is massive enough, towards the end of its life it will be made up of layers of material, with iron group elements in the innermost core surrounded by a shell rich in elements such as silicon, another shell rich in nuclei of carbon, oxygen, neon and magnesium, a helium shell, and an atmosphere of hydrogen and helium. Nuclear burning could be going on in all of these shells at the same time (but not in the inner core) in the last stages of the star's activity – but not for very long.

Astronomers have great confidence in their models of what goes on inside a massive star before it explodes in a final blaze of glory because they were able to study the death throes of just such a star in 1987. That year, a star in the nearby Large Magellanic Cloud, a small satellite galaxy associated with the Milky Way, was seen to explode as a supernova.[5] And the observed

[5] The star is about 160,000 light years away from us so that we saw it by light which left 160,000 years earlier. The supernova explosion actually happened 160,000 years ago, not in 1987.

details of that explosion closely matched the predictions of the behaviour of supernovas made by the computer models. Even better, it turned out that the progenitor star had been photographed in astronomical surveys before it exploded, so we could see what it used to be like.

This combination of theory and observation tells us that the star which exploded began its life only about eleven million years ago, with about eighteen solar masses of material (then, mainly hydrogen and helium). In order to hold itself up against its own weight, it had to produce so much energy that it shone 40,000 times brighter than the Sun, and used up its hydrogen fuel in just ten million years. Then, helium-burning sustained it for about another million years, before it began to run into serious trouble. As the core began to shrink further, the star's collapse was temporarily halted by converting carbon into a mixture of neon, magnesium and oxygen – but only for about 12,000 years. Fusion processes involving neon held the star up for about twelve years, oxygen-burning did the trick for just five years and in the last fling of the fusion process silicon-burning provided enough heat to delay the collapse for about a week – but that was the end of the line.

With no more internal supply of heat, the inner core of the star collapsed suddenly, shrinking down to a ball only a few tens of kilometres across at about a third of the speed of light. As gravitational energy was released, the temperature soared above 10 *billion* degrees, and protons and electrons were forced together to make neutrons. Meanwhile, the outer layers of the star – several solar masses of material – were left with no means of support, and started falling inwards towards the core. The collapsing core itself generated a shock wave inside itself and tried to 'bounce' outwards again, like a golf ball that is squeezed tightly and released; but this shock wave ran into the infalling material and was trying to push away all that mass, the entire outer region of the star. The only reason that the shock succeeded in blowing away the bulk of the original star is that vast numbers of neutrinos produced by all the inverse beta decays in the core ran into it and pushed it on its way.

This gives you some idea of the density of the material inside the shock wave of a supernova. Neutrinos are so reluctant to interact with everyday matter that about 70 billion neutrinos produced by nuclear reactions inside the Sun pass through every square centimetre of the Earth (and you) every second, without being affected. To a neutrino, lead is as transparent as glass is to a photon. But the density in the shock wave inside a supernova is so high (an unimaginable hundred thousand billion times the density of water) that it is as impermeable to neutrinos as a brick wall is to a ping-pong ball.

The result is that the outer parts of the star are shoved away at a speed of about 10,000 kilometres per second, carrying with them huge amounts of heavy elements produced not only by fusion reactions that took place during the lifetime of the star but by reactions forced 'uphill' from the iron valley by these extreme conditions, making elements heavier than iron to add to the mix. It is amazing enough to think that elements such as carbon and oxygen in your own body were made inside ordinary stars, not much bigger than the Sun; it is even more amazing to think that elements such as gold, which many people wear on their fingers, were made not in any ordinary star, but in the cauldron of a supernova, where a single star can briefly shine as brightly as a whole galaxy like the Milky Way.

As the outer layers of the star blew away into space and thinned out, the neutrinos were able to stream through the debris and go on their way at close to the speed of light. One of the most dramatic aspects of the story of the 1987 supernova is that a total of just nineteen neutrinos from the explosion were detected by instruments here on Earth, after taking 160,000 years on their journey from the Large Magellanic Cloud. To put that in perspective, astrophysicists calculate that some 10^{58} neutrinos were produced in the blast. As these spread out uniformly in all directions, at a distance of 160,000 light years there should 'only' have been about 300,000 billion neutrinos from the supernova passing through one of the detectors, a tank containing more than 2,000 tonnes of water deep underground in Japan, where just eleven of them interacted with electrons in such a way that they left a trace. Eight more neutrinos from the supernova were 'seen' by a different detector in the United States[6]. To put the reluctance of neutrinos to interact with everyday matter in a human perspective, about 10 billion neutrinos from the supernova passed through your own body in February 1987 – and you didn't feel a thing!

Tiny though the numbers of detected neutrinos are, they exactly match the expected sensitivity of the detectors to the supernova neutrinos, and this match between theory and observation is one of the greatest triumphs of astrophysical theory and modelling. Not just *astro*physics, either – the whole model of how a supernova works also includes classical physics (in the form of those equations describing temperature and pressure variations inside the star) and nuclear and quantum physics (to describe the interactions which produce the energy that blows it apart). You could say that just about the whole of physics is involved in modelling a supernova, so the success of those models is a triumph for the whole of physics.

A hundred and sixty thousand years before those nineteen neutrinos were detected on Earth, over in the Large Magellanic Cloud, while they and their billions of companions were involved in blowing the outer layers of the star apart, nuclear reactions proceeded frantically in the compressed material just outside the inner core of the star, where temperatures, forced higher by the gravitational energy released during the collapse, soared to 200 billion degrees. All those nuclei that were originally made by adding alpha particles together, and which therefore had equal numbers of protons and neutrons, merged to form huge quantities (about a solar mass) of nickel-56, which has 28 protons and 28 neutrons in each nucleus.

But nickel-56 is unstable, and decays spontaneously, each nucleus spitting out a positron as it converts a proton into a neutron, and itself into a nucleus of cobalt-56. The cobalt-56 itself then decays, in similar fashion, into iron-56, which has 26 protons and 30 neutrons in each nucleus, and is stable. It is the energy released in this process, the decay of radioactive elements

[6] Of course, these detectors were specifically 'looking' for neutrinos produced by interactions here on Earth, or in the Sun; it is sheer good luck that they were up and running at the right time to detect the supernova neutrinos, which were easily distinguished both by their arrival time and their very high energies.

into stable iron, that keeps a supernova shining brightly for weeks after the initial explosion in which the nickel-56 was formed. The unstable nickel-56 was itself built up from the input of gravitational energy during the collapse – the shining light of the star left behind after the explosion is some of that stored-up gravitational energy escaping. About a tenth of the iron made in this way escapes into interstellar space, some to become part of the stuff new stars and planets (and, on at least one planet, the steel in such things as penknives and car bodies) are made of.

Meanwhile, the inner core of the star has collapsed all the way to become a neutron star – and if that inner core still had more than three solar masses of material (which would not be the case for the explosion of so small a star as the one we saw as a supernova in 1987), it would continue collapsing into the ultimate black hole state.

There is another way to make a supernova – slightly less spectacular, but still important in seeding the Galaxy with heavy elements. Many stars are in binary systems, and sometimes one of the stars in such a binary system will have run through its life cycle and become a white dwarf while the other star is still at the red giant stage of its life. In those circumstances, the compact white dwarf can attract matter from the extended atmosphere of its companion gravitationally, slowly gaining mass as it does so – but remember that the maximum mass for a stable white dwarf is only 1.4 times the mass of the Sun. If the white dwarf starts out just below this limit, it can attract enough mass to push it over the edge, triggering collapse to the neutron star state, and releasing almost as much gravitational energy as in the kind of supernova we have just described.

Even if the white dwarf in such a binary system is too small for this to happen, hydrogen from the atmosphere of its companion can accumulate on its surface until there is enough there to trigger a flash of nuclear fusion, briefly making the star shine brightly and sending another load of heavy elements off into space. Unlike a supernova explosion, this lesser kind of outburst – called a nova – does not disrupt the system in which it occurs, so the process can repeat time and again.

All of this recycling of material enriches the interstellar

medium, and provides the raw material for stars like the Sun, planets like the Earth, and people like us. We would not exist if previous generations of stars had not been active in this way. The primordial stuff out of which the earliest stars formed (at least ten billion years ago) contained only 75 per cent hydrogen, 25 per cent helium, and just a trace of the next lightest element, lithium, produced in the Big Bang in which the Universe was born (*see* chapter eleven). The Sun was born, only five billion years ago, out of a cloud of material that was already enriched so much that nearly one atom in a thousand was of an element heavier than lithium.

The oldest stars in our Galaxy have abundances of iron, for example, that are only one hundred-thousandth as great as the abundance of iron in the Sun. But there is still room to improve our understanding of the lives of the stars, since nobody has yet found a star which contains no elements heavier than lithium. The very first stars, in which heavier elements were processed for the first time, seem to have long since faded from the scene, and we cannot know for sure just what it was that started the whole stellar process going. The best guess is that very massive superstars formed when the Universe was young, ran through their life cycles correspondingly quickly, and seeded the clouds from which the Milky Way formed with enough heavy elements to set in motion the processes we have described in this chapter.

This leads us out into the Universe at large, beyond the Milky Way and back in time towards the birth of the Universe. Our story is very nearly complete – but the end turns out to be the beginning, and the large and small turn out to be inextricably linked, like Ouroberos, the snake that bites his own tail.

CHAPTER ELEVEN

THE LARGE AND THE SMALL

The exploration of the Universe at large only began in the 1920s – because, before that time nobody knew that there was a Universe at large. The accepted picture of cosmology was that the Milky Way system *was* the Universe, and that although individual stars might be born and die, the general pattern of the Milky Way was eternal and unchanging. There had been dissenters from this view – as far back as 1755 Immanuel Kant had suggested that some of the fuzzy blobs of light (known as nebulae) seen in the heavens with the aid of telescopes might be other 'island universes' like the Milky Way; but very few people took this idea seriously until well into the twentieth century.

The first step towards an understanding of the relationship between the Milky Way and the Universe at large came when the Milky Way itself was mapped out in some detail for the first time, at the end of the second decade of the twentieth century. The mapper was Harlow Shapley, a young researcher at the Mount Wilson Observatory in California, where he had access to what was then the best astronomical telescope in the world, a new 60-inch diameter reflector. Using this instrument, Shapley was able to analyse the light from Cepheid variables across the Milky Way, and thereby determine their distances. The actual numbers he came up with were slightly too big, because he had not allowed for the way in which dust in space dims the light from distant stars; but it was Shapley who showed, in a series of papers published in 1918 and 1919, that the Sun and Solar System sit far out from the centre of the Milky Way, and do not occupy a special place even in our own island universe.

Because he had overestimated the size of the Milky Way, Shapley thought that the Magellanic Clouds were relatively small systems within the Milky Way, not independent star systems. This encouraged him to think that all the starry nebulae[1] were satellite systems, glorified star clusters in orbit around the Milky Way.

But other astronomers disagreed. As more of these nebulae were discovered and studied, using new telescopes like the 60-inch, and as many of them were seen to have a flattened, disc-shaped appearance with spiral structure, the argument that they must be systems like the Milky Way, at very large distances, gained strength. In 1920, this led to a formal debate, organised by the US National Academy of Sciences in Washington, between Shapley (arguing the case that nebulae were satellites of the Milky Way) and Heber Curtis, of the Lick Observatory, who argued that the spiral nebulae were each major star systems broadly similar to our Milky Way. The debate was inconclusive, but it marked a turning point in cosmology, the moment when the idea of galaxies beyond the Milky Way became respectable. Within ten years, the case argued by Curtis would be proved, with the aid of the latest of the great telescopes on Mount Wilson, the 100-inch reflector.

The proof came from work by Edwin Hubble, who found that the new telescope was just powerful enough for him to be able to pick out Cepheid variables in some of the nearer nebulae – what we would now call galaxies. In order to measure the brightening and dimming of a Cepheid, of course, you need a series of photographs over a period of time. And in order to obtain even one photograph of the required quality Hubble had to spend hours keeping the target locked in the sights of the telescope while the photographic plate was exposed. It took most of two years (1923 and 1924) for Hubble to obtain fifty decent photographs of one particular fuzzy blob of light in the sky (now known to be an irregular galaxy); but the effort was worth it. The

[1] As well as nebulae made of stars, there are other fuzzy blobs, also called nebulae, which are simply clouds of gas inside our own Galaxy. The confusion of names should not be a problem here; in this chapter we are only talking about external nebulae, star systems outside the Milky Way.

Cepheid technique showed that this fuzzy blob was seven times further away from us than the Small Magellanic Cloud, and could not possibly be a satellite of the Milky Way. And at about the same time, Hubble photographed Cepheids in the great nebula in the constellation Andromeda, getting a distance of just under a million light years (this estimate has since been revised up to above two million light years, as our understanding of Cepheids has improved). The Andromeda 'nebula' was proved to be a spiral galaxy very similar to our Milky Way.

At first, the full, shattering implications of these discoveries were hidden. Although it became clear that the Milky Way is just one galaxy among many, because the Cepheid distance scale had been calibrated incorrectly the other galaxies all seemed to be rather closer, and therefore rather smaller, than modern estimates would suggest. Until the 1950s, it seemed that the Milky Way was a giant among galaxies – possibly the largest of them all – even though there are very many other galaxies. But successive improvements in distance measurements have made it clear that the other galaxies are further away, and bigger, than used to be thought. This has reduced our perceived importance in the Universe.

A study in which I was involved in 1997, using observations of Cepheids in many nearby galaxies, made by the Hubble Space Telescope, showed that our Milky Way is indeed an average-sized spiral, compared with all the spirals close enough to have had their distances measured by the Cepheid technique. If anything, our Galaxy is slightly smaller than the average galaxy of its type. It is also estimated that at least a hundred billion galaxies are, in principle, visible to the Hubble Space Telescope, although only a few thousand have been studied in any kind of detail. The Earth orbits an ordinary star which is just one of several hundred billion stars in the Milky Way Galaxy, and the Milky Way Galaxy is just an average spiral, among a hundred billion or more galaxies in the visible Universe. But the most dramatic discovery of all – also made by Hubble – is that the Universe is not eternal and unchanging. It is changing and evolving as time passes.

To a cosmologist, even a galaxy like the Milky Way, containing several hundred billion stars, is no more than a 'test particle' whose behaviour can be used as a guide to how the Universe as a whole is changing. It happens, as Hubble was the first to appreciate, that galaxies come in different varieties. Disc-shaped galaxies like our own Milky Way (flattened by the effects of rotation) make up about 30 per cent of the population of the Universe, while 60 per cent of all galaxies are in the form of ellipticals (shaped, as the name suggests, like a three-dimensional ellipse). Some of the ellipticals are much smaller than our Galaxy, but some are much larger – the largest galaxies in the known Universe are giant ellipticals containing thousands of billions of stars. Apart from spirals and ellipticals, the other 10 per cent are irregular galaxies, starpiles with no distinct shape. To a cosmologist, all of this is mere detail – the behaviour of galaxies that matters most to cosmologists is the way that the galaxies seem to move, revealed by the famous cosmological redshift.

Even before Hubble came on the scene earlier astronomers had noticed that light from many of the 'nebulae' showed a redshift. This was interpreted as a Doppler effect, implying that these objects were rushing away from us through space, but caused great bafflement in the mid-1920s because some of these objects were moving with what seemed at the time extraordinary speeds – 600 kilometres per second, or more.

Measuring redshifts is easy, provided a galaxy is bright enough for its light to be analysed spectroscopically. The great advance Hubble made was not measuring redshifts, but measuring distances – and then putting redshifts and distances together. Hubble's early measurements of distances to galaxies using the Cepheid technique only worked, even with the 100-inch telescope, for a few nearby galaxies. In order to estimate distances to more remote galaxies he had to use a variety of other indications, such as measuring the brightness of exploding stars in those galaxies. It is a reasonable assumption that all novae or supernovae peak at much the same brightness, so this can give you a rough and ready guide to distance. But this kind of technique can only be approximate, which is why all estimates of galaxy distances beyond the immediate neighbourhood of the Milky Way were distinctly vague for many years – really, until the advent of the Hubble Space Telescope at the end of the 1980s.

Nevertheless, even with these difficulties, by the end of the 1920s Hubble and his colleague Milton Humason had found evidence for the most profound discovery in astronomy – perhaps the most profound discovery in science. The redshift of a galaxy is proportional to its distance from us. This means that its 'recession velocity' is proportional to its distance. The entire Universe is expanding, and must have been in a more compact state in the past. It must, indeed, have had a beginning.

Hubble's discovery was so surprising (and, to be honest, initially based on so few galaxy distance measurements) that it might have taken a long time for it to be accepted by his colleagues. After all, *why* should the Universe be expanding? But the answer to that question already existed! Indeed, a complete theory describing the behaviour of the expanding Universe already existed. That theory was the general theory of relativity, completed by Albert Einstein in 1916 and applied to cosmology (by Einstein himself) in 1917.

The general theory of relativity is a theory of space and time, which treats time as the fourth dimension of a space-time continuum, and describes the relationships between space, time and matter in a series of equations. The way to get a feel for this description of the Universe, without wading into the mathematics, is to imagine suppressing two of the four dimensions, and visualising space-time as a stretched, two-dimensional rubber sheet. The effect of the presence of a massive object in space can be pictured by imagining a bowling ball placed on the stretched sheet, where it makes a dent. Small objects moving past the heavy object (marbles rolled across the sheet) follow curved paths, because of the indentation made by the heavy object. This curving of their trajectories is the process that corresponds to the force we call gravity. In extreme cases a very massive, compact object will make a hole right through the stretched sheet – a black hole, with a gravitational field so powerful that nothing, not even light, can escape from it.

Einstein's theory is one of the greatest triumphs of science, and has been tested in many ways. It explains how light from distant stars seems to be bent when it passes by the edge of the Sun (an effect visible during a solar eclipse, when the dazzling light of the Sun itself is blotted out by the Moon), and how black holes with masses millions of times greater than the mass of our Sun swallow up matter and release gravitational energy as they do so, converting it into electromagnetic radiation which pours out into the Universe from the centres of some galaxies, making the so-called quasars. In principle, the general theory describes all of space, and all of time – the entire Universe. Naturally, Einstein wanted to use it to describe the Universe mathematically. But when he tried to do so, in 1917, he found that the equations did not allow for the possibility of a static, unchanging Universe. According to his wonderful new theory, space-time could be expanding, or it could be contracting – but it could not stay the same.

Remember, this was at a time when people still thought that the Milky Way was the entire Universe, and that it was unchanging, in the same way that a forest is unchanging. So Einstein, well aware that 'if it disagrees with experiment then it is wrong' added an extra term to his equations, simply to hold his model universe still. This extra term, the cosmological constant, he later described as being the biggest blunder of his career; but this seems a little harsh, since all he was trying to do was make his model match the best available experimental (that is, observational) evidence of the time. A little over ten years later, with the discovery that the Universe *is* expanding, the cosmological constant could be thrown away. Einstein's original cosmological equations had predicted the discovery of the redshift–distance relationship, even though Einstein himself had not believed it could be true. The theory did indeed agree with observation.

Even better, the theory helps to remove the most obviously puzzling feature of the redshift–distance relationship. Why should everything in the Universe be receding uniformly *from us*? Surely we do not live at the centre of the Universe?

Indeed not. The kind of redshift law that is predicted by the general theory of relativity and that Hubble discovered, with redshift proportional to distance, is the only kind of law (apart from a static universe) which looks the same whichever galaxy you are sitting in. The best way to picture this is to go back to the image of space-time as a stretched rubber sheet, or, in an often used analogy, as the expanding surface of a balloon that is steadily being blown up.

Imagine the surface of the balloon to be scattered with spots of paint, with each paint spot representing a galaxy. As the balloon gets bigger, all of the spots get further apart from one another – *not* because the paint is moving across the surface of the balloon, but because the skin of the balloon (space-time) is stretching, and carrying the spots along for the ride. If the balloon expands so that the distance between two neighbouring spots doubles in size, then the distance between every pair of spots doubles in size. Choosing any spot on the balloon to measure from, a spot that was 1 cm away will now be 2 cm away (seeming to have moved by 1 cm), a spot that was 2 cm away will now be 4 cm away (seeming to have moved twice as fast), and so on. The further away a spot is the faster it will seem to be moving (the bigger its redshift), whichever spot you measure from.

The redshift is *not*, however, a Doppler effect – it is not caused by the galaxies moving through space. It is caused by space itself stretching, which stretches the light from distant galaxies to longer wavelengths on its journey through space. And, just as there is no centre to the expanding surface of the balloon so there is no centre to the expanding Universe. All points in space are on an equal footing (except for local disturbances to space-time caused by the presence of stars, galaxies and so on).

The general theory of relativity also tells us that if we imagine running this expansion backwards in time, the entire visible Universe must have emerged from a mathematical point – a singularity – at a definite time in the past. No physicist believes that the general theory can quite be pushed that far, because quantum effects are certain to become important close to the singularity. But it is clear that the Universe as we know it has emerged from a very dense state, something very close to a singularity, at a definite time in the past. The big questions are – when did this happen, and how?

The 'when' question has been answered by the latest studies of how fast the Universe is expanding, the latest calibrations of the redshift–distance relationship. As in Hubble's day, the key thing is measuring the distance to galaxies – redshifts are easy. Once you have both redshifts and distances, you can calibrate the redshift–distance law, and combine this with Einstein's equations to calculate when the expansion began. The faster the Universe is expanding the less time must have elapsed since the Big Bang for the galaxies to be as far apart as they are today. The expansion rate is measured in terms of a number called Hubble's Constant, or *H*. This number is determined from those measurements of redshifts and distances to galaxies today, and the smaller the value of *H* the bigger the age of the Universe.

There are many different techniques for measuring distances across the Universe today, and happily they all seem to be converging on the same answer. One of the simplest is the one which I used, together with Simon Goodwin and Martin Hendry, jumping off from the technique we used to measure the relative size of our own Galaxy. Since this technique tells us the average size of a spiral galaxy like our own, it is relatively easy to take a couple of thousand spiral galaxies which have known redshifts (but are too far away for their distances to be measured directly by the Cepheid technique) and work out what value for the redshift–distance relationship would be appropriate to ensure that the average size of all those galaxies matches the average size of the nearby spirals.

For many values of the relationship, the distant galaxies will seem to be very far away, and therefore very big in order to look as big as they do from Earth; for another range of values, they will seem to be very close to us, and therefore very small; but for one value of the relationship they will seem to be at just the right distances for their average size to match the size of the Milky Way and its neighbours. This calibration of 'Hubble's law', as it is now known, tells us that the Universe is about 13 billion years old. For the convenience of working with round numbers, many cosmologists use an age of the Universe of 15 billion years, and nobody is likely to argue much with that. What matters is the clear evidence that the Universe did have a beginning, about that long ago – not a billion years ago, not 100 billion years ago, but somewhere around 15 billion years ago.

It is important to stress that this is good science. We have models, based on Einstein's general theory of relativity, and we have observations – not just the way light from galaxies is redshifted but also a weak hiss of radio noise that fills the entire Universe, and is interpreted as the leftover electromagnetic radiation from the fireball in which the Universe was born. Theory and observation agree perfectly. But where does the good science end? We have already said that no scientist believes that the Universe actually began in an infinitely dense singularity – infinities in the equations are a clear sign that the equations are being pushed too far. So how close to the singularity can we probe with the aid of the kind of science that we can do in laboratories here on Earth? The answer may surprise you.

The most extreme conditions of density that physicists can claim to understand completely here on Earth are the conditions that exist in the nucleus of an atom. Protons, neutrons and nuclei are thoroughly understood, and the behaviour of matter at such densities has been probed many times in experiments involving particle accelerators. We have very good models indeed for the behaviour of matter at nuclear densities. Most physicists would go further – they would argue that we have a rather good understanding of the structure within protons and neutrons, at the level of quarks. But let's be cautious, and accept only that we understand completely all the forces at work on the nuclear scale. How far back in time was it when the entire visible Universe that we can see today would have been in a state of nuclear density, as it emerged from the Big Bang?

Taking the hypothetical singularity as 'time zero' – the moment when the Universe came into being – everything we can see today would have been in this nuclear state just one ten-thousandth of a second after the beginning. And everything that has happened since, for the ensuing 13 or 15 billion years of the history of the Universe, can, in principle, be explained by the same laws of physics that have been tried and tested many times in experiments here on Earth.

It is the early phase of the life of the Universe, from about an age of one ten-thousandth of a second onwards, that is usually referred to as the Big Bang. And it is good science, by the criteria we have established – the models agree with the observations.

Just how the Universe got into the state it was in at an 'age' of one ten-thousandth of a second is less clear, but is still the subject of respectable scientific investigation, as we shall see shortly. What is certain is that everything from one ten-thousandth of a second onwards is as well founded as Newton's laws, or Maxwell's equations. It is certainly as well founded as the general theory of relativity, since it uses the general theory of relativity. In round numbers, we can say that the Big Bang lasted for about half a million years – from one ten-thousandth of a second to the time when electromagnetic radiation and matter went their separate ways.

Electromagnetic radiation played a dominant role in the very early Universe, because it was so hot. If you squeeze air into a bicycle pump, it gets hotter; if you allow gas to expand from the nozzle of a spray can, it gets cooler. In the same way, when the Universe was in a very compressed state it was very hot, and we can calculate how hot using Einstein's equations and our observations of the state of the Universe today. Back at the time of nuclear density, the temperature of the Universe was about one thousand billion degrees (Celsius or Kelvin, the difference doesn't matter for such large numbers) and its density was 100,000 billion times the density of water. In this fireball, individual photons carried so much energy that particle–antiparticle pairs (such as a proton and an antiproton) were constantly being created out of pure energy, and constantly annihilating one another to release electromagnetic energy (photons) once again. At first there were equal numbers of protons and neutrons around (and also very many electron–positron pairs); but as the Universe expanded and cooled, several things happened.

First, the temperature dropped sufficiently for photons to be unable to make either proton–antiproton or neutron–antineutron pairs any more, so the numbers of these particles froze out. Because of a tiny imbalance in the laws of physics, which scarcely shows up at the temperatures we can achieve on Earth today, the numbers of particles and antiparticles had not quite been perfectly matched up in pairs in the original fireball, and for every billion antiprotons there were a billion and one protons, while for every billion antineutrons there were a billion and one neutrons. All the billions of pairs of particles and antiparticles annihilated one another, making high-energy photons, leaving the odd one in a billion left over. Everything we see in the Universe today is made from this tiny fraction of leftover material, and there are a billion photons in the background radiation for every nucleon in the Universe.

Those photons were, at that stage, far from finished as an influence on matter. Although they no longer had the energy to make new protons and neutrons, they interacted violently with the remaining particles, encouraging the decay of neutrons into protons, with individual neutrons each spitting out an electron and an antineutrino, and becoming a proton.

As time passed, the proportion of neutrons in the primordial soup declined. Just a second after the beginning, the temperature of the Universe was down to ten billion degrees, the density was only 380,000 times the density of water, and there were only 24 neutrons left for every 76 protons. By 14 seconds after time zero, there were only 17 neutrons left for every 83 protons, but the pace of change was slowing dramatically as the Universe cooled. About three minutes after the beginning, the temperature of the Universe was down to about a billion degrees, 70 times the temperature at the heart of the Sun today, and there were still 14 neutrons around for every 86 protons.

If the remaining neutrons had been left as free particles, within a few more minutes they would all have decayed naturally into protons (even without being bombarded by super-energetic photons). But the Universe was now cool enough for some of those protons and neutrons to stick together, building up stable nuclei of helium-4 (alpha particles). By the time the Universe was four minutes old, all the remaining neutrons had been locked up in this way, giving a mixture of 74 per cent hydrogen nuclei (protons) and 26 per cent helium nuclei (alpha particles) as the Universe entered the next phase of its life.

All of this story is based on our understanding of the laws of physics determined from experiments in laboratories here on Earth, plus the observed fact that the Universe is expanding, and the cosmological equations of the general theory of relativity. One of the most striking successes of the cosmological models is that they predict this mixture (26 per cent helium and 74 per cent hydrogen), which is just the mix seen in the oldest stars, made of primordial material.

For hundreds of thousands of years after the helium formed, the Universe expanded quietly. It was still far too hot for electrons to become attached to those nuclei and form atoms, so the electrons roamed freely among the nuclei, forming a plasma, and all the charged particles, (electrons and nuclei alike) interacted with the electromagnetic radiation that still filled the Universe. The photons richoceted from one charged particle to the next in a crazy zig-zag dance, just as they do inside the Sun today. The next dramatic change – the end of the Big Bang – happened between about 300,000 and 500,000 years after the beginning, when the entire Universe cooled to about the temperature at the surface of the Sun today, a mere 6,000 degrees.[2] At that sort of temperature, electrons and nuclei can stick together to form electrically neutral atoms; and electrically neutral atoms scarcely interact with electromagnetic radiation at all. Suddenly, the Universe became transparent. All the billions of photons from the primordial fireball were free to stream through space uninterrupted, while the atoms could begin to clump together, undisturbed by the radiation, to form clouds of gas, collapsing under their own gravity to form the first stars, and the galaxies.

The radiation, which had a temperature of about a billion degrees when the helium was made and a temperature of 6,000 degrees when it decoupled from matter half a million years after the beginning, should, according to the models, have cooled all the way down to about 3 K today (that is, *minus* 270 degrees Celsius). The second great triumph of the Big Bang model is that exactly this sort of radiation is detected coming from all directions in space – the cosmic microwave background radiation.

[2] It's worth pausing for thought here, to try to put the Big Bang in perspective. This is about the first time in the story of the Big Bang that we have encountered a temperature we can understand in human terms. We have, after all, all seen the Sun, and felt its warmth, at a distance of 150 million kilometres. At the time matter and radiation decoupled, the entire Universe was just like the surface of the Sun today. And in order to get to this fantastically hot state (by human standards), it had already been cooling for half a million years.

This radiation, first detected in the 1960s, is almost perfectly uniform; but in the 1990s, first the COBE satellite and then other detectors discovered tiny variations in the temperature of the radiation from different parts of the sky. These ripples are like fossils, marks imprinted on the radiation the last time it interacted with matter, 500,000 years after the beginning. So they tell us about the way matter itself was distributed across the Universe, the last time it interacted with the radiation, just before it started collapsing to form stars, galaxies and clusters of galaxies. Those irregularities are exactly the right size to have been the seeds from which the structure we see in the Universe today, from galaxies upward, would have grown by gravitational collapse.

Where did those ripples come from in the first place? What seeded the structure into the Big Bang itself? Now, we have to move on to slightly more speculative ground, and bring the ideas of particle physics into cosmology. As is usually the case with a developing field of science, there are several models which are offered to explain what went on in the beginning. For simplicity, I shall give you just one, my favourite, as a guide to the way science thinks that it all began. The variations on the theme present the same sort of broad picture, although the details may differ.

The big question, in philosophy and religion as well as science, is why the Universe should exist at all – why is there something rather than nothing? The answer, from a combination of quantum physics and the general theory of relativity, is that the Universe itself may, in a sense, be 'nothing'. Specifically, the total energy of the Universe may be zero.

This is a startling thought, since we know there are hundreds of billions of galaxies, each containing hundreds of billions of stars, and Einstein taught us that mass is a highly concentrated form of stored energy. But Einstein also taught us that gravitation (warped space-time) is a form of energy – and the bizarre thing is that a gravitational field actually stores *negative* energy. It is quite likely that all of the mass energy in the Universe is precisely cancelled out by all of its gravitational energy. The Universe may be nothing more – or less – than the ultimate quantum fluctuation.

To get this idea of negative gravitational energy clear, think of something much smaller than the Universe – a star, say. Imagine taking a star apart into its component pieces, and spreading them out as far apart as possible, to infinity. It doesn't matter whether you think of the component parts as atoms, or nuclei, or quarks – the argument is still the same. Remember that the force of gravity obeys an inverse square law. So the force between two particles that are infinitely far apart is one divided by infinity squared, which is very definitely zero. The amount of gravitational energy associated with a *set* of particles that are infinitely spaced out is also zero.

Now imagine allowing the particles to fall together. Of course, if they are infinitely far apart this will never happen, but we can imagine giving them a nudge to set the ball rolling.[3] One of the most important things we know about gravity is that when a collection of particles falls together under the influence of gravity, energy is released – the particles get hot. This, after all, is how the Sun and stars got hot enough inside for nuclear reactions to begin in the first place. Without this release of gravitational energy in a collapsing object we would not be here. At the particle level, what is going on is that energy from the gravitational field is being converted into energy of motion (kinetic energy), so the particles fall faster. When they collide with one another this kinetic energy of the individual particles shows up as heat. But this is the strange thing about the gravitational field. We started out with zero energy. Then, as the particles fell together, the gravitational field gave up energy to the particles. Which means that it now has less than zero energy – for all real objects in the real Universe, the energy associated with the gravitational field is negative!

It gets better. If you imagine the gravitational collapse continuing all the way to a point – a singularity like the one that the general theory of relativity says was associated with the birth of the Universe – the total amount of energy released by the

[3] All of this is an example of what astronomers call a 'hand-waving' argument, but it makes the point we want to make. If the equivalent calculations are carried through properly, using the general theory of relativity, they give the same surprising answer.

gravitational field is exactly the same as the mass–energy of all the particles that have fallen together. If the mass of a star like the Sun were concentrated at a point, the negative energy of its gravitational field would exactly balance the positive mass–energy of all its matter. It would have zero energy overall. The same argument applies to the whole Universe. If it was *born* at a point, instead of collapsing to a point, then the vast amount of mass–energy in all of the matter in the Universe put together could be precisely balanced by the equally vast negative energy of its overall gravitational field, giving zero energy overall.

Of course, just as quantum effects blur the singularity at the birth of the Universe, so they would blur the singularity associated with a collapse of this kind. But the kind of quantum effects that do this blurring involve uncertainty – a true singularity cannot exist in quantum physics, because there is no such thing as a precisely determined amount of energy at a precisely determined point in space-time.

Quantum uncertainty, as we discussed in chapter three, allows 'empty space' to be alive with packets of energy that appear out of nothing at all, and disappear within the time limit set by the quantum rules. The less energy involved the more time the 'virtual' energy packet, a quantum fluctuation, can exist before the Universe notices, and it has to disappear. Thus a quantum fluctuation which had precisely zero energy overall could, as far as the quantum world is concerned, last for ever!

In the 1970s, a few cosmologists toyed with the idea that the entire Universe might be the product of a quantum fluctuation of this kind, something that just popped out of the vacuum for no reason at all, simply because this is not forbidden by the laws of physics. The way this would happen, according to those quantum models, would not be at a singularity. The quantum rules say that there is a smallest possible interval of time, called the Planck time, and a greatest possible value for density (the Planck density), as well as a smallest possible interval of distance (the Planck length). Putting all these limits together, the Universe would have been born in a state when it was already 10^{-43} seconds (the Planck time) old, and would have had an initial density of 10^{94} grams per cubic centimetre. The entire visible Universe might have originated from such a Planck particle, about 10^{-33} cm (the Planck length) across, one hundredth of a billionth of a billionth of the size of a proton.

The big snag with this idea, though, was that it seemed obvious, in the 1970s, that such a Planck particle would have a very short lifetime, whatever the quantum rules allowed. It would, after all, have an absolutely enormously strong gravitational field, and this gravity would, it seemed, crush it back out of existence as soon as it appeared.

But cosmology was transformed in the 1980s by input from the world of particle physics, and especially from the investigation of the way the forces of nature combine at high energies. As we discussed in chapter three, the best models of the way particles and fields interact tell us that at very high energies the four forces of nature we know today would have been united in one superforce. That would have been the situation in the Planck particle (or particles) at the birth of the Universe. But immediately, at the scale of the Planck length, gravity would have split apart from the other forces, and gone its own way; the other forces would have followed suit very rapidly. According to the models, this splitting-apart of the forces of nature would have converted some of the available energy into an enormous outward push, forcing the seed of the Universe to expand dramatically in a tiny fraction of a second, and completely overwhelming (for that split-second) the tendency of the gravitational field to make the quantum seed collapse.

The energy released in this way is often likened to the energy, known as latent heat, which is released when water vapour condenses to make liquid water. As the water goes from a high-energy state to a low-energy state latent heat is released. This is the driving force of a hurricane, for example, where so much heat is released by condensing water that vast swirling convection currents of air are produced. The winds in a hurricane swirl because of the rotation of the Earth; but the Universe does not rotate. The birth of the Universe was like a cosmic hurricane, but with the winds all blowing in one direction – outwards.

This idea has become known as inflation, and the key feature of inflation is that it was an exponential process. The way the

Universe expands today is more or less linear – if two galaxies were situated so that the distance between them doubled in 5 billion years, then it would double again in the next 5 billion years (actually, it wouldn't *quite* double in the second 5 billion years, because the present expansion is slowing down under the influence of gravity).

During exponential expansion, though, if the distance between two particles doubles in one second, it quadruples in the next second, increases eightfold in the third second, and so on. Exponential expansion very quickly runs away with itself. In the very early Universe, the doubling time was about 10^{-34} seconds, which means that in the space of just 10^{-32} seconds the size of the early Universe doubled a hundred times. This was enough to take a seed one hundredth of a billionth of a billionth of the size of a proton and inflate it to about the size of a grapefruit, all before the Universe was even 10^{-31} seconds old. At that point the forces of nature had all split apart, and the inflation stopped. Even so, the grapefruit-sized Universe (containing all the mass–energy of the entire visible Universe today) was left expanding so rapidly, in the Big Bang, that gravity could not halt the expansion for hundreds of billions of years.[4]

Astronomers are excited by the idea of inflation, because it provides a model which explains how the Big Bang began. Particle physicists are excited by the idea of inflation, because it provides them with a test of their own theories, at energies far greater than anything that can be achieved here on Earth. And both camps are delighted because inflation, like all good scientific

[4] There is an intriguing variation on this idea, which we cannot resist mentioning. We have suggested that a quantum fluctuation that was the seed of the entire Universe could have appeared out of nothing at all, in the quantum equivalent of a singularity. But such quantum singularities are precisely what is thought to form at the centres of black holes, where matter collapses to a point. It is seriously suggested by some scientists that this collapse towards a singularity may be reversed by quantum processes, including inflation, bouncing (or being shunted sideways, in some sense) to create new universes that expand away into their own versions of space and time. Every black hole in our Universe may be the gateway to another universe; and our Universe may have been formed by the collapse of a black hole in another universe, making an infinite sea of bubble universes in the vastness of space and time, and removing the idea of a unique origin to everything. And since the total energy involved both in the collapse and the expansion is zero, it doesn't matter how much (or how little) mass goes into the collapsing black hole – it can still make a full-sized universe on the other side.

theories, makes predictions which can be tested, in this case by observation of the Universe at large. By and large, inflation passes these tests – although there are still many things we do not understand about the birth of the Universe.

The most impressive test that this whole package of ideas has passed so far concerns those ripples in the background radiation that we have already mentioned. The ripples we see today were imprinted when the Universe was about half a million years old, and show that matter (in the form of clouds of hot gas) was distributed at that time in just the right way to collapse and form stars and galaxies within the expanding Universe. According to inflation theory, when the Universe visible to us today was packed into a volume only 10^{-25} cm across (a hundred million times bigger than the Planck length) 'ordinary' quantum fluctuations, like the ones we described earlier in this book, should have produced tiny ripples in the structure of the Universe. As well as expanding everything else during inflation, the process of exponential growth should have taken those quantum ripples and stretched them, stamping their imprint across the Universe, in the form of variations in the density of matter from place to place.

The kind of ripples in the background radiation seen by COBE and other detectors exactly correspond to the kind of ripples that would be produced by blowing up quantum fluctuations in this way. If the models are correct, without the primordial quantum fluctuation there would have been no Universe at all; and without these secondary quantum fluctuations there would be no people around to puzzle over the origin of the Universe, because everything would have been smooth and uniform, so no stars would have formed. No other model yet devised can explain why the Universe is both very uniform overall (thanks to inflation), but contains exactly the right kind of ripples needed to make galaxies and clusters of galaxies (also thanks to inflation).

This is an example of the way inflation explains something that had already been noticed – the fact that we exist. Can it make a real, testable prediction? One that can actually be tested in laboratories here on Earth? Yes. One of the key requirements of all the variations on this theme is that the amount of matter in the Universe should indeed balance the negative energy of the gravitational field. This precise balance is required in order for the Universe to expand out all the way to infinity, and hover there, without recollapsing – the time-reversed mirror-image of our original description of a cloud of infinitely dispersed particles collapsing to a point. In the language of the general theory of relativity, this means that space-time must be flat overall, with no pronounced curvatures.[5]

In more everyday language, it means that the expansion we see going on today will not carry on for ever, but nor will it be reversed by gravity to make the Universe recollapse. This infinite hovering only happens if there is a critical amount of matter in the Universe, which would correspond to about a hundred times more matter than all the bright stuff we see in the form of stars and galaxies. In other words, it is a prediction of inflation that most of the mass of the Universe is in the form of dark matter.

This is fine by the astronomers, since they already know that there is more to the Universe than meets the eye. From studies of the way galaxies rotate and move through space, it is clear that they are being tugged on by unseen matter – the astronomical evidence shows that there is at least ten times as much dark matter as there is bright stuff, and maybe a hundred times as much.

But the calculations of how matter was processed in the Big Bang, which we outlined earlier in this chapter, set limits on how much of this matter (bright or dark) can be in the form of protons and neutrons (together forming nuclei of baryonic matter, the kind of stuff we are made of). The processes that went on in the Big Bang – interactions involving protons, neutrons, electrons and neutrinos – are crucially affected by the density of this material during the first few seconds, at the time when these interactions are taking place. In order to come out with the right mix of 26 per cent helium and 74 per cent hydrogen, you cannot have more than a few per cent of the critical amount of matter today in the form of baryons. There can be other kinds of matter,

[5] There are subtle variations on the theme which allow the Universe to be very nearly, but not quite, flat. This does not affect the discussion below.

provided that it does not get involved in the nuclear fusion reactions that made helium when the Universe was young, but you can't make the Universe flat using baryons alone.

This is fine by the particle physicists, because their standard theories of particles and fields – the very same models that predict inflation in the first place – also predict the existence of different kinds of particles from anything we have yet seen, non-baryonic matter. These particles have never been detected because, according to the models, they do not feel the strong force, or the electromagnetic force (they do not even feel the weak force). They only interact with the rest of the Universe by gravity – in effect, they got cut off from everything else as soon as gravity split off from the other forces, right back at the birth of the Universe. Even though these particles have yet to be detected, the models predict what kind of masses and other properties they should have, and they have been given names, such as photinos.

More generally, this whole class of object is referred to as Weakly Interacting Massive Particles, or WIMPs – for the obvious reason that they have mass, but do not interact very strongly with everyday matter. Although many different kinds of WIMP can exist in principle, and would have been made in profusion in the energetic conditions of the very early Universe, it is likely that only the lightest of the WIMPs (possibly the photino) would be stable, and that all the other WIMPs would have decayed into this lightest state very early on in the history of the Universe.

Of course, what both astronomers and particle physicists would like is a way to detect these particles directly, and that may happen within the next few years. Now that their presence is suspected, it ought to be relatively easy to catch a WIMP, since the masses prescribed for them are typically about the mass of a proton – and also because the models suggest that we are swimming in a sea of WIMPs.

If WIMPs have masses in the same sort of range as the mass of a proton or a neutron, and yet there is up to a hundred times as much dark matter in the Universe as there is bright stuff, there must be up to a hundred times as many WIMPs as there are baryons. Unlike the baryons, which are concentrated in stars and galaxies, the WIMPs would be spread out more evenly through the Universe (there is indirect evidence for this, in the form of the patterns made by clusters of galaxies on the sky; computer simulations show that the observed pattern of galaxies can only have formed if the bright baryonic matter is embedded in a more uniform sea of cold, dark matter).

Simple calculations suggest that there may be as many as 10,000 WIMPs in each cubic metre of the space around us – not just the *empty* space, but also the space occupied by ordinary matter, including the air that you breathe, the solid Earth, and the heart of the Sun (the figure looks slightly less impressive, though, when you remember that at a temperature of zero degrees Celsius, every cubic centimetre of air at sea level contains 45 billion billion molecules). These WIMPs would speed through a laboratory at speeds of a couple of hundred kilometres a second, like a swarm of cosmic bees, and most of them would pass through any human-sized object without noticing it was there.

Out of all these WIMPs, just a few will happen to strike the nucleus of an atom and will recoil from it (remember that, as Rutherford discovered, an atom is mostly empty space, with just a tiny central nucleus). In a kilogram of material (any material) there are about 10^{27} baryons (a billion, billion, billion baryons). Somewhere between a few and a few hundred WIMPs (depending on their exact properties, which are not yet known) will be involved in a nuclear collision inside a kilogram of material each day; and this is the basis of the WIMP detectors now becoming operational.

If a WIMP actually collides with the nucleus of an atom, that nucleus is going to notice. But under everyday conditions nuclei are constantly being buffeted – by cosmic rays from space, by thermal motion, and so on. This background noise has to be eliminated if the effects of collisions with WIMPs are to be discerned. So WIMP detectors involve taking blocks of very pure material, which are kept sealed away from outside influences at very low temperatures, in laboratories at the bottom of mine shafts, or in tunnels beneath mountains. There, they are cooled to a few degrees above the absolute zero of temperature (a few Kelvin), and watched by sensitive detectors to see if they are disturbed by the impact of a WIMP.

Detectors with the sensitivity to record the impact of WIMPs with just the kind of properties predicted by the Grand Unified Theories of particle physics (the same properties that are required by astronomers to explain the nature of the dark matter in the Universe) are now starting to become operational, and there have already been a couple of claims of a detection, although these have largely been dismissed as premature by scientists outside the teams making the claims.

The decisive experiments will be carried out early in the twenty-first century; then, if WIMPs do exist, they really will be seen individually. For the first time, we will have direct experimental contact with particles representing 99 per cent of the mass of the Universe. But negative results from these experiments will mean decisively that the WIMPs do not exist – that the models are wrong.

That would be a pity, because the most exciting thing about the whole WIMP saga is that the properties of the WIMPs predicted by the particle models are *exactly* the properties of the particles required by cosmologists to provide the dark matter in the Universe. It is the most extraordinary agreement between branches of science which seemed originally to be going in opposite directions from one another, one delving down into the world of the very small, the other looking outward into the world of the very large.

If WIMPs are detected in the next few years, and they do have the predicted properties, this match between the large and the small will be the greatest vindication of the whole scientific method. But if they are not detected, we will have to think again – not about the scientific method, but about this particular set of models. Either way, you now have a ringside seat at the testing of one of the most important and far-reaching ideas that science has ever come up with. Whatever the outcome of that test, though, it will do nothing to alter the truth of the most fundamental and important thing that science has to teach us. No matter how beautiful the whole model may be, no matter how naturally it all seems to hang together now, if it disagrees with experiment, then it is wrong.

FURTHER READING

Dawkins, Richard, *River out of Eden* (Weidenfeld & Nicolson, London, 1995).

Eddington, Arthur, *The Nature of the Physical World* (Folcroft Library Editions, Folcroft, Pennsylvania, 1935).

Emiliani, Cesare, *The Scientific Companion* (Wiley, New York, 2nd edn, 1995).

Feynman, Richard, *The Character of Physical Law* (Penguin, London, 1992).

Feynman, Richard, *Six Easy Pieces* (Addison-Wesley, Boston, 1995).

Feynman, Richard, *Six Not So Easy Pieces* (Penguin, London, 1998).

Fortey, Richard, *Life: An unauthorised biography* (HarperCollins, London, 1997).

Fritzsch, Harald, *Quarks* (Allen Lane, London, 1983).

Gribbin, John, *Companion to the Cosmos* (Weidenfeld & Nicolson, London, and Little, Brown, New York, 1996).

Gribbin, John, *Q is for Quantum* (Weidenfeld & Nicolson, London, 1998).

Gribbin, John and Mary, *Richard Feynman: A life in science* (Viking, London, and Dutton, New York, 1997).

Gribbin, Mary and John, *Being Human* (Phoenix, London, 1995).

Mason, Stephen, *Chemical Evolution* (Oxford UP, 1992).

Murdin, Paul and Lesley, *Supernovae* (Cambridge UP, 1985).

Scott, Andrew, *Molecular Machinery* (Blackwell, 1989).

Trefil, James, *From Atoms to Quarks* (Scribner's, New York, 1980).

Watson, James, *The Double Helix* (Critical edition, ed. Gunther Stent, Weidenfeld & Nicolson, 1981).

Weiner, Jonathan, *The Beak of the Finch* (Jonathan Cape, London, 1994).

INDEX